THE AUTHOR

C.L. Prajapati (*b.* 1949) has done his post-graduation in Physics, History and Indology & Museology and doctorate in conservation of Paper Documents. He is presently working as Scientific Officer in the School of Archival Studies, National Archives of India, a pioneer educational institution in Afro-Asian countries offering various courses in Archival discipline including care and conservation of Books, Manuscripts and Archives of Professional level. He had started his career in 1972.

Prajapati has contributed articles, which were published in various reputed journals and participated in various conferences/seminars in the field of conservation of documentary wealth. He was actively engaged in preparing reading materials relating to conservation, brought out by NAI entitled *Pralekhon, Pandulipion Tatha Dusprapya. Pustkon Ka Abhirakshan Aur Pratisanskar* (Hindi), *Elements of Records Management and Conservation* (English).

The author has conducted several practical workshops on Conservation of Documentary Heritage and Records Management. He has also authored a book entitled, *Archivo-Library Materials Their Enemies and Need of First Phase Conservation.*

CONSERVATION OF DOCUMENTS

PROBLEMS AND SOLUTIONS

—POLICY PERSPECTIVES

C. L. Prajapati

MITTAL PUBLICATIONS
NEW DELHI-110059 [INDIA]

First Edition 2005

ISBN 81-7099-922-7

Showroom:

4594/9, Daryaganj,
New Delhi - 110002
Phone: 23250398

Published and printed by Krishan Mittal for Mittal Publications,
A-110, Mohan Garden, New Delhi-110059, India.
Tel.: 25351493, 25351976 *Telefax*: 91-11-25351521
E-Mail : mittalp@ndf.vsnl.net.in *Gram* : Mittalbux
website : www.mittalpublications.com

Dedicated to

My Wife Rajani,
Daughters Manju and Meena
and Sons Manoj, Ashutosh and Ravinder

ACKNOWLEDGEMENTS

There are many directions from where invaluable help has come to complete this work.

Intellectual help and moral support extended to the present author by scholars, well-wishers and professionals may not be conventionally measured. But my *Guru*, Dr. Susmita Pandey, Assistant Professor, Prachya Niketan, Bhopal, the supervisor of my dissertation work, did not only devoted her precious time for this work, but also, in the process, broadened the "conservation" perspective of the author whose debt may not be reciprocated in any way.

The author is also indebted to Dr. S. Sen Gupta, Officer on Special Duty, National Archives of India, for extending help and motivating from time to time to carry on the work and also making of valuable suggessions.

The author acknowledges as well his sincere gratitude to Dr. S. Sarkar, Director General of Archives for granting him permission for doing work, extending moral support and guidance from time to time.

The author is grateful, to Prof. Dr. A.K. Das of National Museum, Deemed University, New Delhi and Dr. P. Mehra of National Archives of India for devoting their valuable time for sympathetic guidance.

The author would like to extend his hearty thanks to his friends and colleagues, Shri V.K. Juneja and Shri N.D. Mankey of National Archives Regional Centre Bhopal, whose untiring efforts made it possible to bring out the work.

The author expresses his sincere thanks to his colleagues Shri Rajesh Verma, Shri Jagmohan Singh, Shri A.K. Dey, Smt. Santosh Tyagi and Shri Shanti Sarkar for their motivation, and supply of material. Gratitude is also expressed for skilful assistance from Shri. Hemant Kumar and Shri Asarfi Ram, of National Archives of India.

Last but not the least, author gives his heartiest and warmest thanks to his wife, Smt. Rajni Prajapati who has led the torch and motivated him to proceed steady to the goal.

C. L. PRAJAPATI

PREFACE

Conservation of paper based record is the subject most talked about today. The job of conserving information or knowledge has been one of the important aspects of all the civilized societies of the past, and the means and ways adopted for the purpose have been changing from time to time. In the societies, where people were unaware about the art of writing, the system adopted for the purpose of conserving information was different from that which was adopted in the post era of invention of art of writing. In the subsequent periods with the advancement of time as all kinds of awareness increased and the organic materials such as papyrus, leather, barks leaves and paper were adopted as vehicle, the job of conservation was once again reshaped, in order to face the challenges, within the framework of available scientific know-how of conservation. If we peep through the past, it becomes clear that the job of conservation almost followed a set pattern so far as the system is concerned but restorative means went on changing with the introduction of new materials as carrier of information. In modern times, with the introduction of paper, invention of dyes and their use in manufacturing of ink, industrialization and change in local atmospheric conditions, the people associated with Archives, Libraries and alike institutions were once again compelled to rethink over the subject and define conservation with a new vision.

The Indian sub-continent, in the post independence period has been very vigilant and active in all aspects of conservation of paper heritage, housed in Archives, Libraries and allied organizations. Various means and ways, were found out to answer the local problems correctly. Various techniques were invented to fulfil the requirements of the subcontinent. But the result of such toil is lying in fragments and needs a systematic study, consolidation and analysis, so that it may be truly used by the people entrusted with the job, of conversation. The work as such, I hope, will not only help them in facing challenges in conservation of documentary wealth but also help in planning effective conservation programme in the face of challenges ahead. This idea inspired me to take up the work of study under the subject title, "Conservation of Documents, Problems and Solutions – Policy Perspective ". For systematic study the entire work is divided in 9 chapters.

The first chapter, talks about the concept of conservation in Indian societies of the past, its growth, development and new trends in modern times.

The knowledge about nature of a document being essential part in the conservation of documentary heritage, an effort has been made in the second chapter to deal with this aspect, nature of components, paper and ink, in making of a document. Without understanding the materials in making of paper, ink and their components, it is difficult to understand the nature of a document and without this aspect one can not decide the course of action to be taken for chemical treatment and reinforcement. Third chapter put forth the relevant information on the agencies so far known responsible for decay, damage and spoilage of documents while the fourth chapter deals with the means and ways found suitable for preservation of the documentary wealth against the enemies. The unique character of the chapter is that it covers the herbal based insecticides and pesticides for curative conservation. Commercial insecticides and fumigants are not free from side effects on records as well as on human beings, hence the talk on the materials of plant origin to control insects and fungus, makes the study significant for safer use and also to plan systematic study of conservation for future.

Paper being organic in nature, whatever care is made, one day it is bound to become weak and also possible to be damaged because of handling, use, man made calamities and intrinsic dangers. The fifth chapter, therefore, dwells on restorative measures of damaged records irrespective of any reasons and salvaging them. It covers not only the techniques for chemical treatment and reinforcement but also the materials required for carrying out restorative programme and the parameters under which it should be carried out for effective results.

The sixth chapter highlights the problems faced by the people in running conservation programme, since conservation of documentary heritage is a delicate work. Delicate in the sense, if without understanding the problem properly, a document is subjected to chemical treatment and reinforcement, every possibility is that instead coming out any good after it, the document may suffer irrecoverable loss. Such incidents, if happen, may bring about a permanent loss to the information world and need to be avoided at any cost. This is the reason why it is said that every document is an individual problem. Hence this chapter has its own merit.

The Indian conservation today is so advanced that it works as an advisory force within the country. It also advises Afro-Asian countries. How, gradually, it was achieved during the fifty years of independence and what is the role of Indian Historical Records Commission in its development, has been discussed in the seventh chapter of the present Text.

The eighth chapter, throws light on the conservation of manuscripts housed in Satras of Majuli. It is the wealth on sanchi pat, a new support, the world has not seen much of.

As a conservation scientist what I understand is that for effective conservation, a National Policy is essential. What should be the major constituent of the policy, they have been covered in ninth chapter. Besides, to make the subject matter more meaningful, perceptible and effective, relevant information relating to subject has been appended and figures, plates and charts have been included as and where deemed necessary.

C. L. PRAJAPATI

CONTENTS

LIST OF TABLES

LIST OF CHARTS

LIST OF CHARTS

LIST OF FIGURES

LIST OF PLATES

1

INTRODUCTION TO CONSERVATION AND ITS DEVELOPMENT IN MODERN INDIA

Conservation is a term which is being used widely today in various fields such as conservation of forests, conservation of wild life, conservation of soil, conservation of energy and momentum, besides conservation of documentary and cultural heritage. Different etymons are being used to define conservation in respective fields but the soul of conservation that is to keep the things intact safe in idea, quantity, quality, shape and size for the use of posterity, is found common everywhere.

The concept and use of conservation, is an age old tradition. The sole idea behind conservation is to ensure stabilization and protection of records against dangers and also correction of damaged wealth. The sense of ensuring security against every agent endangering life is a phenomenon which concerns every organism, microscopically small or monumentally large. In the very idea of security lies the concept of conservation. The radius of conservation sphere starting from self, diverges covering the entire world so far as Indian society is concerned. The concept of safety and security towards life against endangering agencies is nothing but a nucleus for conservation. Thus the antiquity of conservation may be said to be as old as the evolution of life.

The conservation in II phase being mostly limited to self life, may be termed as conceptual conservation. The last phase of conceptual conservation, however, specially in Indian society, is that period when slogans like "Vasudhaive Kutumbakam" and "Ahimsa Parmodharmah", emerged advocating welfare for all in a wide perspective. The material conservation, however, is not less than 4400 years old as it started much before the time of erection of pyramids of Egypt, which are only the oldest embodiment of material conservation as the objective behind their erection was to preserve the dead bodies of Egyptian Kings.

> "The oldest example of material conservation in world history is Pyramids of Egypt. The remarkable story behind the Pyramids is that people of Egyptian royal family had a custom to bury the dead bodies of their Kings in pyramids, with the belief that after thousand years or so the dead Kings will come back to life and they will look for their bodies. While narrating this fact, the royal Priest claims the knowledge of preserving dead bodies of their Kings, a secret to the royal family. The Pyramids have been dated back to 4400 years. No doubt the Pyramids are famous for their glory but at the same time they may be treated the mile stone in the history of material conservation because no other source than Pyramids is available in world literature which throws light upon the scientific preservation of organic materials. Though, it is not known whether the people of royal family could keep dead bodies preserved, for thousand years or so, yet the remarkable fact which comes out of the story is that people of that time had already started experimentation for preservation of organic materials. Thus the antiquity of material conservation may be fixed to the period prior to 4400 years, that is the age of Pyramids, as the thought of preserving dead things of organic nature would have appeared much before this age."

Indian literary and mythological sources are replete with examples of manifesting the idea of preserving memories through the ages. Few of them are construction of stupas, tombs and other monuments. The sole purpose behind the erection of stupas was to preserve the memories of lord Buddha and his teachings. After the invention of art of writing, for permanent preservation of the information, inscribing over the stone (pillars and caves) was made. Stone inscriptions of Ashokan period and the period of Guptas, (Prayag Prasasti) are the well known example of this kind. Besides, the Gurukul system of education, during ancient period, was also started for the purpose of educating people, disseminating information to various parts of the country and also to keep the information preserved from generations to generations. These facts converge to the point that conservation as a system has been in vogue, in one form or the other, from time immemorial. But, in modern age, word conservation appeared perhaps for the fist time in the scientific arena as conservation of energy, and momentum.

To define the term conservation, applied for the documentary heritage, it is a method of keeping documentary heritage, safe from decay, damage and spoilage and to stabilise and strengthen them as and when necessity arises.

In the early parlance the world, "Preservation" was used as synonym of conservation. But as the time advanced, the activities for protecting the materials also advanced and the term preservation fell short of communicating the sense of activities performed towards keeping cultural heritage safe. As such, it was replaced by the word conservation. According to Dr. G.M. Cunha, "Library conservation is two-fold; Preventive measures and restorative measures after damage has occurred.

So far activities carried out towards conservation, in the field of documentary heritage, the glossary of Keeping Archives, published by Australian Society of Archives, Sydney, in 1987, states that it is composed of two things and they are the Preventive Conservation and the Restorative Conservation. The two terms are inadequate to project the entire activities carried out in conservation today and can be classified as under which are known as the elements of conservation.

(i) The Examination, Research and Development.

(ii) The Preventive Conservation.

(iii) The Curative Conservation

(iv) The Restorative Conservation.

(v) The Duplicative Conservation.

The preventive and curative activities are jointly known as prereservation. The term preservation is defined as the adoption to appropriate prophylactic and prospective measures for maintaining the documentary heritage in as good condition as possible and to prolong its life to the maximum possible extent. In common practice, preservation deals with the maintenance aspect.

The activities imparting strength to the decayed and damaged specimens and to restore them to the possible level of rejuvenation is, however, termed as restorative conservation. Reformating or copying activities of documentary wealth is defined as duplicative conservation. The examination, in general, is common to all the elements of conservation.

National Archives of India is the sole representative of Paper Conservation and whatever development achieved in the apex organisation, they reflect the developments brought out on the subject in the country. The beginning of conservation in Modern India, begun as past as 1860, in the form of suggestion by Hugh Sandeman, a civil auditor, for the first time, for weeding out records (created by the agencies of Government of India) of ephemeral nature and shifting those of permanent value to a centralised muniment room for safe preservation. No doubt, the execution of suggestion was unduly delayed and could be implemented only in 1891, when G.W. Forrest (1891-99) took over as an Officer-in-charge of Records of the Government of India. Thus the modern National Archives (then Imperial Record Department) started to roll on. Subsequently, various achievements were bagged which could be understood by systematic study in two parts; *i.e.*

(i) Development of Conservation in pre-Independence period.

(ii) Conservation performed in the post-Independence period (after 1947).

During the tenure of Forrest, the records of Government of India lying in various creating agencies, were centralised for proper preservation. The period of Samuel Charles hill (1899-1903) is witnessed for introducing systems of flattening of folded records, repair of damaged documents using tracing paper and their placement in file covers for preservation. The other significant developments, in pre-independence period, are the search for new material for the purpose of reinforcement of damaged records. (1911-12), the research and development activity conducted by Sudburrough and Mehta to find out the cause of embrittlement of paper (1920) and establishment of a Research Laboratory in the Imperial Record Department in 1940.

The guiding force perhaps behind establishment of Conservation Research Laboratory was the necessity of finding causes relating to the nature of materials used for creation of records and also to materials employed for reinforcement of fragile records.

Archives conservation, which was in its infancy initially had now (after setting laboratory) started to upsurge finding out right means and ways setting conservation on appropriate tracks. It can be manifested from the three major achievements of 1911-12, 1920 and 1940 all related to research and development. The infrastructure developed in pre-independence period had started bearing fruits in the post-independence period which is obvious from the achievements of the time.

Developments of Conservation in Post-Independence Period

The conservation activities before 1947 were mostly based on the techniques being deployed in the Public Records Office, London. But, after 1947, when personnel of National Archives of India were got trained, the new era of expedient means and ways suiting the local conditions got an opening and achievements in all the elements of conservation were recorded.

To carry out scientific examination of documents and the materials to be used for preservation and restoration of documents, laboratory was lacking the adequate facilities and therefore, in 1948, its modernisation scheme was chalked out. Under the scheme, Ph meter, electric muffle unit were procured (1948). Subsequently, Petti Foggers (humidifiers) bursting strength tester, ultra-violet lamp, etc. were procured in 1949. Having these facilities, the work of scientific examination of materials required for restoration of fragile documents got a starting. Samples of hand-made paper, received from Research Laboratory, Hyderabad and Khadi & Village Industry

Commission, Bombay and sample of Chiffon cloth received from Government Silk Weaving Factory, Jammu & Kashmir were tested. Thus, the deck was cleared for using materials made through indigenous sources.

To ensure preventive norms in the stack areas, the department's proposal for air-conditioning of the stack areas was approved by the Government of India in 1947. However, the facility due to one reason or the other could only be availed in 1970.

It may be recalled that for curative conservation materials like insecticides, pesticides and fumigants (commercial) were introduced in 1937. But, in 1939, first time, Naphthalene was introduced as insect repellent. Subsequently, Paradichloro--benzene in 1941 and Killoptera (a mixture of carbon-terra-chloride and Ethylene Dichloride) in 1949 were adopted for fumigation of insect infected records. Moreover, in 1951, for mass fumigation, the most effective technique, through vacuum fumigation was introduced. With the introduction of this technique, it would become possible now to annihilate insects with their eggs from huge records at a time within a short duration of 5-6 hours.

The restorative conservation technique was mainly traditional and manual in India in pre-independence period. It was mechanised partially, in the post-independence era. As sources speak, important mechanisation in this field took place between 1947 and 1955. Equipments of important nature such as steam platen lamination machine (1947 from USA), a mass air cleaning unit (1951)[+i], rotary ironer[+ii] and Pettifogger[+iii] etc. were acquired for the purpose during the period.

The acquisition of hydraulic flat bed laminator, revolutionised the job of reinforcement of decayed and damaged records. It also systematised the work of reinforcing documents written with water soluble ink which was an up-hill task till then.

Development of new indigenous techniques, giving right answer to the problems for reinforcement of documents is not less important than other achievements. Solvent lamination (1951) special full pasting technique for repair of multilayered documents (1992) and techniques for storage of Palm Leaf

Manuscripts (1994) are the few techniques developed indigenously. Besides, fabrication of fumigation chamber in 1990 and formulation of glue paste are the other achievements of this period.

To ensure effective conservation the role of standards is very significant. Standards like recommendations relating to the design of building for libraries (IS: 1553-1989), archives [(IS: 2663-1989), code for anti Termite measures for building, (IS: 6813 (Pt. II) 1981] and their amendments No. 3, 4, and 5 respectively in 1991, 1993 and 1995 and standard specifications for ball point pen ink (IS: 5805-1993) were got formulated by the Bureau of Indian Standards.

For dissemination of technical know-how and to make trained personnel available for various custodial agencies and libraries in the field of conservation, two courses; one of Professional Level and another of Sub-professional Level, *viz.* the Care and Conservation of Books, Manuscripts and Archives; and the Serving and Repair of Records were introduced in 1980 and 1973 respectively in the School of Archival Studies, National Archives of India. This step, fulfilled the growing demand of custodial agencies for trained and expert restorators and conservationists.

Duplication of records is considered an effective aid to conservation. Though, the technique of microfilming, a device of duplication on reduced scale, came into practice with the introduction of flow cameras in the second decade of the 20th century, it was adopted by the archivo-library world for security and preservation of records in 1930. This technique for the first time entered India through National Archives in 1943 and in the same year one microstat equipment along with accessories was installed. Subsequently, one microfilm reader from U.S.A. (1945) was received and order for procurement of 52 microfilm storage cabinets and a micrographic camera was placed in 1947. Hectic efforts were made during 1948-52 to improve the micrographic system by acquiring equipment and materials, like Microfile Recordak Model C & D Cameras, Recordak Portable Projector, Recordak Library Film Reader, Speedgraphic Camera with a flash syncroniser, Placko Economy Drier and Washers, Recordak Enlarger, Omega Enlarger, Film

Rewinders, Dry Mounting Press and 600 Rolls of Microfilms, etc.

By 1971, Indian Conservation had come up as an apex advisory body. The reputation and the fame of Indian conservation crossed the boundaries of our country. In this very year (1971) on the invitation from Malaysia, experts from this Department were deputed to train the staff of Malaysian National Archives. Subsequently, in 1981 the experts were deputed to set-up a repair unit in Malaysian National Archives. In recent past, in 1997 an expert from this Department was deputed to Kenya for organising workshop on conservation.

Estimating the achievements of conservation during past 50 years, it can be said that it has made alround success and did not remain confined to the boundaries of the country but leaped abroad.

The talk on new trends of conservation is of significance in to day's perspective. It emphasises on the use of quality materials for creation of records and preventive measures in departmental record rooms as well as in Archival repositories so that the documentary heritage may be kept intact avoiding their restoration for centuries. Without preserving information through the ages, it would have not been possible to make such a tremendous progress and write a different history by human being.

REFERENCES

1. Australian Society of Archives, *Keeping Archives* (1987), Sydney.
2. Prajapati, C.L., "Genesis of Conservation and its Development in Modern India", *The Indian Archives,* Vol. XLVI, No. 1-2, January-December, 1997, p. 53.
3. Prajapati C.L., "New Trends in Conservation", *Proceedings of the Meeting of National Committee on Archives,* Shimla, 1993.
4. Prajapati, C.L., "Conservation of Manuscripts in Tawang Monastery", *The Indian Archives,* Vol. XXXIX, No. 1, January-June, 1990, pp. 23-28.
5. Prajapati, C.L., *Archivo-Library Materials their Enemies and Need of First Phase Conservation.*
6. *Conservation of Traditional Records, Paper and Allied Materials,* p. 40.

+(i) "A cleaning unit is the mechanical system employed for removing accumulated dust from the records on mass scale. The significance of the system, is that it sucks the dust particles, instead pushing them to float in the atmosphere, and force them to accumulate in a chamber where from they are collected and ousted.

+(ii) Rotary ironer is a mechanical device for flattering of folded document.

+(iii) The Pettifogger is also known as humidifier. It is used for separation of blocked mass of documents and also for studying the effect of humidity on organic materials.

2

NATURE OF COMPONENTS OF DOCUMENTS AND THEIR CHANGING CHARACTER WITH THE PASSAGE OF TIME

To conserve cultural and documentary wealth, it is essential to be aware about the components of a document. The two major components, the vehicle and the ink, which themselves are made of various ingredients, are known to their individual character. They behave differently in varied conditions of climate. They are nutrients for many organisms. Moreover, illustrated manuscripts have a third ingredient made of colours, dyes and pigments besides the two referred to above.

The constituents of paper, ink and colours, illustrated manuscript and miniature, more or less, have been changing from time to time. Without the knowledge of these factors, it is difficult to apprehend the cause of their decay. This subject may be discussed under the following three heads:

1. Paper—a dominant carrier of information.
2. Ink.
3. Colours.

Paper—A Dominant Carrier of Information

Support has been of paramount importance, for depiction of

ideas and recording of information since invention of art of writing. The support, the discovery of art of writing or depiction of ideas and the *Kalam* (pen) are triplet. After discovery of art of depiction, various information carriers have been deployed by the people of different ages but paper is today the most influential and important support in all countries whether democratic or totalitarian, bureaucratic or autocratic for educating and influencing public opinion on all matters, political, economic, social and cultural. Numerous questions relating to paper such as the name of first paper inventor, the materials first used in making of paper, country where first paper was produced and eventualities of its nomenclature are to be answered with authentic proof by the scholars. Some Indian conservators have, however, attempted to answer the questions but they are not the final words about it.

About the date and place of first paper, Y.P. Kathpalia has mentioned that the first paper was made in 105 A.D. in China. But there is doubt about the date and the land where first paper took birth. The following facts in this connection are well to be considered. According to Nearcos, commander of Alexander Sinkander's Army, people of Punjab region used to prepare paper from cotton. Nearcos, after the invasion by Alexander against India in 327 B.C., stayed for sometime in Punjab and studied the activities of local people and accounted them in his personal diary. Genesis of that was published in "Indica". Nearcos writes, "Indian were manufacturing paper out of cotton by poundig". G. Buhler & others in work, Indian Palaeography and Dr. T.V. Mahalingam, professor of Ancient Indian History University of Madras in his book Early South Indian Palaeography have supported this view. Maxmooler, a famous Sanskrit scholar of Germany, has also mentioned in his famous account, "History of Ancient Sanskrit Literature" that people of India used to prepare paper from wet-cotton. Dr. Satyendra believes that first paper was produced in India. Besides, in Jain religious book "Yajna Valkya" Smirti, word "Pata" (paper) has been used as writing material. Greek Ambassador Megasthenes (303-299 B.C.) in Mauryan court had described how Indian used paper for writing horoscopes and almanacs.

The above accounts of scholars converge to the fact that

first paper was produced much before Christ and its place of birth was no other part of the world but Asian land and probably India.

The art of paper making seems routed to China, Central Asia, and then other parts of Asian lands with the expansion of Buddhism during the period of Great Hindu Emperor Ashoka. This art was acquired by the Japanese and through them spread to other parts of the world.

Alexander Soteriou, in her text, "Gift of Conquerors, Handpaper making in India", has apprehended that the craft may have entered India by as many as four routes; North-west through Sind, Himalayan North and by Chinese influence into Bengal. The apprehension put forward by the author, actually belongs to a very later age, 8th and 9th century A.D. when Indian crafts had almost deserted and disappeared due to political instability and foreign invasions after the decline of Gupta period. The craft of paper making which was inceptionally Indian, re-entered the Indian region in the 8th century from China. But, it does not mean that people of India were not aware about the art of paper making prior to it.

The stone inscriptions and tablets recovered from the site of Indus Valley civilisation are evidences to the fact that people of this civilisation were aware about the art of writing and depiction though it was symbolic. Further, the stone inscriptions of Ashokan period, followed by Indus valley civilisation, are known for their artistic writing. The style of writing deployed for inscribing, itself tells that it could not have been achieved in a year or two. But it would have taken years to attain the style. Secondly, the inscribing on stone or any hard surface is more difficult than the writing on the surface by chalk or quill pen etc. From the facts, it can easily be deduced that art of inscription, specially on the hard surface, is the second generation writing followed by surface writing. Besides the textile industry in Mauryan age had attained the high order of international reputation. It is evident from the female effigies carved on the stone pillars forming gates on the four directions of the stupa of Sanchi, in modern Madhya Pradesh.

The presence of fibre in the relevant remains of the Indus Valley sites and knowledge of the art of writing among the people of this civilisation, use of textile and riches of carved inscriptions and mature writing style during Mauryan period, compells one to think the possibility of employing support materials of organic nature other than stone. But due to paucity of authentic records, it is not possible to show exactly as to when such organic material was adopted as writing support in India. However, before the Britishers established their business in the country, paper was being used in India during the Mohamedan period. Probably the paper was rag made and manufactured in indigenous-cottage industry. Almost it can be said that paper making in ancient and medieval period in the country was cottage industry. After the Britishers set their feet on the land, the use of machine made paper was also introduced.

According to a monograph on paper making in the Bombay Presidency in 1900 AD., two kinds of paper were being produced: (1) Handmade Paper, and (2) Machine made paper. Variety of handmade paper was prepared chiefly from jute or old gunny bags and from old waste paper. With the establishment of Paper Mill, in the second half of the nineteenth century, the production of machine made paper got a start in the country.

Hand Made Paper

Out of various known and unknown places where paper industry was active some are Kanpur, Mathura, Sanganer (Jaipur), Kota, Delhi, Shahabad, Patna, Kashmir, Khambat, Ahmedabad, Daulatabad (Kagajpura), Safrabad and Khoshundo. These are known for variety of hand-made paper, from the unknown past until recently. Besides, Junnar in Poona district, Erandol in east Khandesh, Nasik, Bagalkot in Bijapur district, Allahabad, Gokak and Talikot are well known places where paper making was a prosperous trade in medieval time.

The paper making craft in Sanganer, a township 10 km. away south of Jaipur city, was so advanced that even weight and type of paper was fixed and had got the name Ishwar Shahi Paper. Ensuring smoothness of paper was categorised as Do-Moharia, one and half Moharia and simply Moharia. The families associated with the craft of the region are still known as Kagdi or Kagji. Dr. Satyendra states that Sanganeri paper was durable and permanent. During the period of erstwhile ruler of Jaipur state, in the seventeenth century the paper was adopted for making ledger books, stamps and creating official records.

Ahmedabad was one of the oldest homes of handmade paper industry even in 1848 as is evident from the following extract. To the north of the city were seen the ruins of old vats for soaking jute rags till 1908 A.D., hence too, the name of the locality Kagdi Pith, says R.T.F., Kirk, an ICS Officer. Further, he says that similar vats are found in Paldhi Kochral, a village in the further side of the Sabarmati river. Ahmedabadi paper was famous for its toughness and whiteness. It was mainly used for providing account books, letters, deeds and book binding. Besides, over most of the Bombay Presidency (1878), Ahmedabadi paper was supplied to different parts of Central India.

> "Thirty years ago (1848) about 800 men and boys were daily employed in the paper works. The workers do not trade in paper. They are generally employed by rich Muslims chiefly of the Sunni Bohra Sect. The employer supplied the materials, chiefly old bags and sacking, much of it brought by Vangaras from Marwar and mixed with pieces of damaged European paper. Under foreign competition paper making has to late years declined. At present (1878) Ahmedabad paper is used chiefly in Govt. Vernacular registers, in Native States and by Native traders, whose ways of book keeping and book binding require tough and close grained paper. Besides, over most of the Bombay Presidency, Ahmedabad paper goes to different parts of Central India. (From *Bombay Gazetteer* Vol. IV, Ahmedabad)"

The production of handmade paper, in 1908, had been confined to Ahmedabad city alone. The sharp decline in industry took place because of European machine made paper. At that time, the craft was being pursued by about one hundred families, who had been forced to work long hours on low wages to earn their living because of poverty.

Junnar in Pune was well known for the paper it produced. The craft has survived till 1908 when it support about one hundred workmen. There were as many as 50 factories in working order.

Erandol, in East Khandesh, was famous for "Erandoli Kagd". Away from the town, on the banks of the Ajnani River there was a place known as Kagadpura. The Houses of the place were fitted with vats and plinths of massive stone and its enormous cement floors (chabutaras, 400 years old) were laid down in the open air by the river-side where the paper pitch was washed and dried. See plates 2.1, 2.2, 2.3, 2.4 and 2.5.

Paper making in Nasik was set up in 1800 A.D. by a Thakur of Brahma Kshatriya, Balaji, Abaji, who bought a colony of Muslim in-habited Kagdis from Roje near Aurangabad. Once upon a time 50 factories were functioning to the north-east of the town, which was then called Kagdipura. In 1885, only 5 to those fifty remained. By 1908, the industry was extinct.

Gokak and Belgaum, by 1908, had abandoned their craft of paper making and adopted weaving. At Bagalkot in Bijapur district, the craft existed but practically died after 1908.

Sajikhar (a black mineral) chiefly carbonate of soda obtained from Bikaner) and quicklime (chuna) were mixed with pulp as filling material for making paper.

Books written on various types of handmade paper such as Mughalia, Kashmiri, Arwal, Sahebkhani, Khambati, Ahmedabadi, Doltabadi, Nanakhambati, Velaor Gajia and Nana Muragi are still available in various libraries of India. The tradition of writing on handmade paper in the country continued even after the introduction and expansion of European machine made paper. The handmade paper of books and volumes still looks young, fresh and strong even after centuries while machine made paper looks dull and weak only within a few years.

In the post independence period, handmade paper industry was recognised as village industry under the Khadi and Village Industries 1957. It generally employs textile fibres derived from rags, gunny bags, cotton linter and other waste material, for making paper. The paper has higher average strength. High cellulose contents of quality, restricted chemical treatment and use of fillers give the paper permanent and durable character. Today, the production of paper is 5000 tonnes per annum. Because of its distinctive properties, the paper is employed for special purposes and it is durable. A variety of handmade paper produced by KVIC is used, in Archieves for full-pasting, guarding, half margin repair, inlaying and also as end paper. The paper is preferred for its qualities of permanence and durability. It has hardly any substitute. The use of handmade paper for archival purposes upholds the proverb, "Ek Anar sau Bimar".

A question flashes in the mind if the materials like paper were used in ancient times, why the existence of such old documents is not reported from any part of the world? The simple answer to the question is the unconducive climate of the region, political instability and attempts of invaders to destroy and demolish the riches of the culture of the country. Moreover, patronage and creation of records and Mss. in limited number are also responsible for their disappearance irrespective of other reasons. The incident of putting the library of Nalanda Vishwavidyala into fire may be quoted as an example. The incident took place when a Chinese traveller Huensong was visiting the country. According to the account of the traveller, the library material had been burning for six months. Though no account of library collection is available, yet it can be said with certainty that the most of library holdings would have organic character and presence of materials like paper in the collections cannot be ruled out.

Development of Paper Industry in Modern India

No doubt, the paper making by manual means was a very old craft of Indians. But, the question is when machine made paper first time started in India? Mr. Y.K. Rao, the author of a book, *The Paper Industry in India, Status and Prospects* has mentioned that paper industry in India started as early as 1832 with the setting up of the first four drienier type machines. However, the commercial production of paper could not be effected before 1880, unless the four mills: (i) The Upper India Couper Paper Mill (1979), (ii) The Titagarh Paper Mills (1882), (iii) The Deccan Paper Mills (1889), and (iv) The Bengal Paper Mills (1887) were set up. What were the raw materials and principal reducing techniques deployed in the afore-mentioned Mills for moulding paper are the subject of further investigation. However, the Indian Paper Pulp Co. Ltd., Naihati, which was established in 1918 during 1st World War, for domestic paper production was based entirely on the utilisation of bamboo as basic raw material. By the time of 1950, India had established 17 paper mills with a capacity of 1,37,000 tonnes and production about 1,16,000 tonnes per annum. The production of paper till the time was confined to common varieties of paper and paper-board.

The paper industry in the post independence period, grew with fast speed and by 1998, approximately 380 paper mills, large and small with varying capacity (Producing Paper 10,000 to 20,000 tonnes) were established. The list of paper mills is indicated in appendix 1.

The Raw Materials for Paper Making

Various raw materials have been used through the ages for manufacturing paper. According to B.B. Lal, a well known Archaeologist of India, linen, cotton and hemp rags were used for making paper in ancient days. Subsequently, straw, jute, bamboo, flax, ramic and woods were also adopted for the purpose. It is point worthy to mention that all above materials used in ancient, medieval and even modern paper industry are from plant origin. Alexander Soteriou, in her book *Gift of Conquerors Hand Paper Making in India* has listed the plants of about hundreds of varieties whose fibres were employed for making paper by hand through the ages. As the sphere of users enlarged and paper technology advanced, the raw materials from other origins to produce cheap and quality paper were tried by the paper industries and paper from fur, hair, silk and wool (animal origin), asbestos (mineral origin), glass fibre was produced. Today paper grass farming has been introduced to produce raw material for making paper.

The principal constituent of cotton, linen and hemp is alpha cellulose which is the most stable material. Alpha cellulose is found associated with beeta and gama celluloses, lignin, resins, pigements, waxes, fats and minerals in the materials of plant origin. Alpha-cellulose having high molecular weight, escapes dissolution in caustic soda (NaOH) of 17.5 per cent strength at 20°C in normal atmospheric pressure while two other celluloses dissolve. Quality of alpha cellulose fibre and length confers strength to the paper. It is a polymer of monomer Beeta glucose having a composition $C_6H_{10}O_5$. Alpha cellulose, a chief constituent of paper though abundantly available in all plants and grasses but it does not exist in free state. It is found in form of bundles. In cotton wool, however, it exists almost in free state and in pure form. For quality paper, separation of alpha-cellulose is essential from the impurities associated with it.

A straight fibre of alpha-cellulose is made when millions of Beeta glucose units link through glycositic bond as shown below. The length of fibre so formed is so small that it cannot be seen by unaided eye. Ratio of hydrogen and oxygen in cellulose is 2:1.

glycositic bond -o-o-o-o-o-o-o←Beeta glucose unit (monomer)

Cellulose fiber

The various kinds of raw materials broadly being used in the modern paper mills in India, may be grouped into three:

(i) Forest Based Raw Materials.

(ii) Agro Based Raw Materials.

(iii) Others.

Salai (Bosevellia, Serrata) eucalyptus and bamboo are the some species which are found in Indian forests suitable for producing paper, mostly newsprint. Other species of forest origin for the purpose of producing paper could be seen in the portion dealing with the newsprint.

Straw and bagasse constitute the major agro based and secondary raw materials of promising alternatives for paper industry in India. Indian forests are lacking woody raw materials for producing paper of quality. As such, the straws from wheat and rice are considered potential raw base offering opportunity to meet the local requirement of paper industry. These are quite different, chemically and morphologically from wood. These are preferred due to environmental concerns and economy.

Straws usually contain less cellulose and legnin but hemi-cellulose and ash (Mostly silicon compounds). See table 2.1.

An experimental study on upgradation of wheat straw and rice straw for pulping and paper making by Sushma Mohajan

Table 2.1 : Comparative Chemical Composition of Wheat Straw, Rice Straw & of Woods

	Lignin%	*Cellulose%*	*Pentosan %*	*Ash%*
Wheat Straw	16-20	29-35	26-32	4-9
Rice Straw	12-16	28-36	23-28	15-20
Hard Woods	23-30	30-49	19-26	Less than 1

reveals the fact that paper produced from these straws utilising upgraded raw (free from nodes and leaves) soda pulping technique and unbleached stocks brings about appreciable improvement in paper properties without any loss in pulp yield.

Table 2.2 : Improvement in Physical Properties of Paper Produced from upgraded Wheat Straw and Rice Straw

Percentage Improved Physical Properties in

	Tear Strength	*Tensile Strength*	*Bursting Strength*	*Brightness*
Wheat Straw	2.7%	42.13%	35.11%	10 Units
Rice Straw	19.2%	17.43%	32.98%	4 Units

The studies of Bhandari K.S. and Srivastava Anita, on pulp and paper making from straws have shown that washing of wheat and rice straw brings about quantitative as well as qualitative improvement in the pulp. This improvement is more apparent in case of rice straw than wheat straw.

Bagasse

Bagasse is considered potential raw material for paper making of either kind (writing and newsprint.) It could be a promising substitute of forest raw material provided its use as a fuel in sugar mill boilers could be suspended. The fibrous residue left after sugarcane is technically known as bagasse. India is the world's largest producer of sugar cane, producing more than 150 million tonnes per annum. It takes an edge over other raw materials due to low Kappa number and excellent bleach ability leading to lower pollution.

The different pulping techniques for producing cultural and newsprint from bagasse are adopted. Chemical or semi-chemical technique is employed for manufacturing cultural paper. In the conventional chemical process in Indian mills, de-pithed bagasse and alkali are fed into horizontal continuous digesters while in semi-chemical pulping, bagasse is cooked in continuous digesters. It has low lignin contents and high yield. Recovery of chemicals from bagasse black liquor has unique problem due to short fibre length, high silica content and high viscosity of the liquor. The problems, however, have been sorted out now to some extent.

The depleting forest based woody raw materials, has forced paper industry to look for suitable alternatives. Bagasse is giving the promising response for the future, concludes Rajesh, K.S. and Rao Mohan N.R. in paper studies.

More recently, the recycled paper has been introduced in the country. Waste paper, is the raw for producing this quality of paper product. So far as strength and permanence of paper, the studies of Chhaya Sharma, N.J. Rao and S.P. Singh reveal the fact that it is weaker. The bonding and other properties of pulp fibres are changed during recycling. The properties of kraft pulp seems to be decreased while that of mechanical pulp opposite to it. However, refining, addition of chemicals, alkali and engyme treatment, bring about improvement in properties of recycled paper.

Moulding of Paper

Any raw material used for making of paper cannot be moulded into paper unless it is converted into pulp. As such, the reduction of raw material into pulp is the first important step in the art of paper making. The slurry made of fibres and water is known as pulp while the process of reduction of raw materials to fibrous state making slurry is called process of pulping.

In ancient days the process of pulping was manual. Mr. Y.P. Kathpalia in this connection has stated that the early papers were produced by stamping or beating hemp, linen and cotton rags in mortars with water until a smooth paste of fibres was obtained. In early times, perhaps, cotton wool and purest form of cellulose available in the nature was utilised in paper industry for moulding of paper. Its use as raw to paper involved no much labour in fibrillation. With a little treatment, it could be made into paper. After cotton wool white rags were the second preferred material for production of quality paper. In the subsequent period, with the dawn of industrialisation, paper making process took a turn and the use of bleaching agents and chemicals for whitening coloured rags and for digestioning them was started.

Rags collected from various levels of human habitats, such as villages, towns and cities, reached the modern paper mills where accepted lot was subjected for dusting and washing. Latter was carried out in big containers bearing holes to allow dirty water drain out of the rags. Subsequently the lot was left in water for week or so for fermentation. Sometimes, for the purpose of softening the rags, the lot was boiled in caustic soda (NaOH) for 10-12 hours. The treated bulk was thus washed in cold water and then turned down in small trucks for sorting of impurities like rubbers and buttons etc. So sorted lot now was subjected for reduction in breaking engines and cleaned with oil of vitriol (sulphurous acid) a bleaching agent known to the people that time for converting coloured pulp white. Then the mass was turned over to beating machine where fibrilation and hydrolysis of pulp took place.

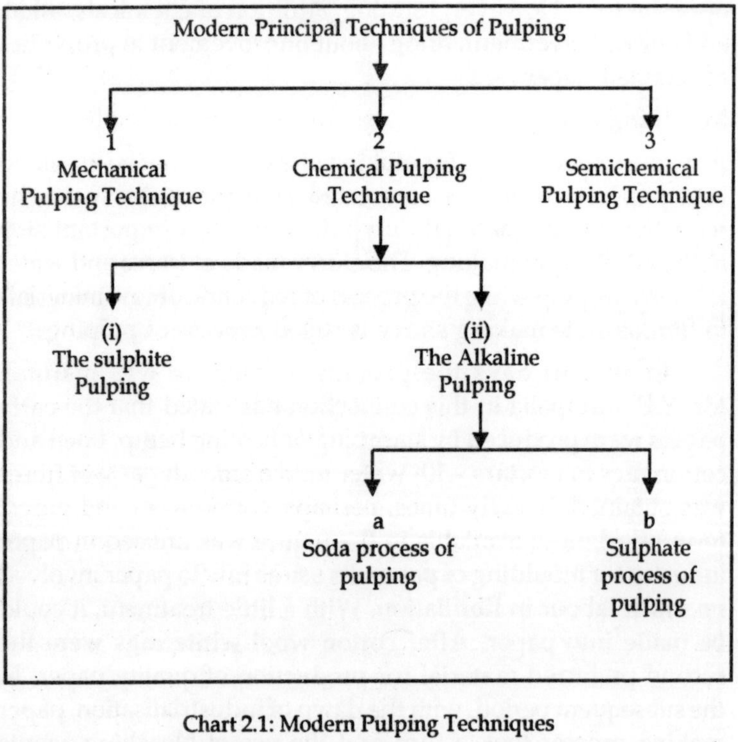

Chart 2.1: Modern Pulping Techniques

The third phase of paper technology seems to be marked by the use of materials of plant origin other than the cotton rags and hemp. Owing to the rising standard of education and its expansion to the common people, the demand of paper rapidly increased leading to the short supply of raw materials like cotton rags to paper industry. Consequently, the search for new raw materials for producing paper started. The attempts of Mathi and Coops, first time claimed the success in manufacturing paper from straw in 1800 A.D. Subsequently, various types of wood (hard & soft) and grasses were tried for the purpose with success. The technique of reducing raw materials was now entirely changed from manual to highly mechanical.

The principal techniques for reducing raw materials of plant origin to pulp are two i.e. (1) Mechanical Pulping, (2) Chemical Pulping. Mechanical Pulping involves the wet grinding of wood into fibrous mass by means of a large revolving grinding stone. The pulp so prepared is known ground wood pulp which differs from chemical pulp. Mechanical pulp contains all the ingredients of wood including lignin and fibres which do not exist as individual entities but into the form of bundles of fibres.

Mechanical pulp is, therefore, much poorer in strength and durability than chemical pulp. The mechanical pulp is much cheaper than chemical pulp.

Before World War II, the world's annual production of mechanical pulp had reached the colossal figure of approximately 10 million tonnes. The major portion of the pulp was used for manufacturing of newsprint. The total production of this paper was nearly 8 million tonnes in 1937.

Chemical Pulping

Chemists Watt and Burger are said the father of Chemical pulping. They authord a new chapter in the history of paper technology by introducing chemical pulping techniques. Because of this invention it could be possible to separate lignin, (a principal unwanted constituent in paper causing damage to it) and alpha cellulose from each other. Both the principal chemical pulping techniques; the sulphite and the alkaline involves the use of cooking liquors, heat and pressure for shoftening or digestion of raw materials. The cooking of hard wood is easier than that of soft one.

The cooking liquor in sulphite pulping technique is a mixture of sulphur dioxide, sulfurous acid and calcium bisulphite. Since the nature of this liquor is acidic, it is also known as acid cooking agent. The cooking of chipped wood is completed in acid resistant bricklined steel digesters or in stainless steel digesters.

The soda and the sulphate process are the two major techniques of alkaline pulping. The main cooking liquor in soda process is sodium hydroxide while that in sulphate the mixture of soda and sodium sulphite (Na_2S). The previous process is generally deployed for digestion of hard wood and latter for soft wood. Digestion of soft wood requires drastic treatment and a long terms cooking of 6-7 hours. Prior to subjecting the wood for digestion, they are chipped properly and then filled in digesters.

More reduction of raw materials into pulp and moulding that into paper does not produce a quality paper suitable for all purpose. As such, to produce white paper with smooth and glassy surface the pulp so produced is subjected for treatments like bleaching, fibre preparation, fibre bonding, filling and loading and sizing.

Paper manufactured from unbleached stock is coloured and not white. As such, bleaching of pulp is an important activity of modern paper mills to produce white paper. Excess and drastic bleaching, however, causes chemical degradation of cellulose and consequently results in poor grade of paper. Therefore, bleaching conditions must be chosen to hold degradation to the minimum. Chlorine and hypo-chlorite are principal bleaching agents. Other bleaching agents used in smaller quantities are chlorine dioxide, peroxide and chlorite.

Paper made out of the bleached pulp but of un-prepared fibres is fluffy, porous and poor in strength, whereas paper prepared from beaten stock, which is a fibre preparation activity, is strong, dense and rattle. Therefore, fibre preparation which involves beating. crushing or rubbing and brushing is an important activity to produce a quality paper. For writing papers, the stock must be well beaten to produce a hard rattle paper. The beating determines the character of a paper and is, therefore, of significance to its durability i.e. its strength.

Filling and Loading

The addition of minerals to pulp stock before moulding it into paper is termed as filling and loading. Use of materials in appropriate proportion, improves the optical and physical characteristics of paper. The use of loading or mineral fibres provides dimensional stability to papers, improves opacity. brightness and receptivity of the surface for printing.

Some important filling agents are clay, talc (agalite and astestine) oxide of titanium and zinc, calcium carbonate, calcium sulphite (gypsum). zinc sulphite and silica.

Excessive loading results in loss of rigidity, abrasiveness on the surface and decreasing of bonding strength of paper, therefore, needs controlled use.

Sizing

Sizing of paper is essential to improve behaviour of paper towards fluids and vapours. Papers resistant to liquids are called sized papers. Sizing binds the fibre and prevents ink from feathering on paper and also imparts hardening. The basis of modern sizing methods is the intimate coating of the individual fibres with a material which is insoluble and repellent towards water. In case of drawing paper, it prevents the rapid absorption of colour, paints and improves finish and strength.

For sizing of paper, two principal processes viz. (i) internal sizing, and (ii) surface sizing are deployed. Sizing agents for internal sizing are mixed with fibrous furnish for uniform distribution of fibres and agents while in surface sizing materials are applied over the surface of moulded sheets of paper. This is performed in size tub, size press or on the calendars.

Starch, yeast of mutton and cattle in water medium were used in early days as internal sizing agents. In the eighties, the pairings from the tanners and curries were recommended for the purpose by parchment makers.

Four principal techniques of surface sizing are; alum-rosin, silicate, starch and gelatine.

Alum was used as sizing materials in Europe in the seventies. Subsequently, in eighties as an economy measure, some paper makers started to add alum rosin size to the pulp during beating operation. Rosin is a solid residue of gum of juices of coniferous trees. By adding alum to rosin solution they react with each other forming a complex hydrophobic compound aluminium resinate which is precipitated on the paper fiber rendering it water repellent. This method of sizing is most effective below PH.5. Rosin was first time introduced in about 1800 A.D.

Alum being acidic in nature, causes embrittleness to paper and stains brown with passage of time. Hence for permanent and durable paper regulation of alum quantity in paper is a must. On exposure to light and air, rosin changes to a crystalline substance resulting in gradual reduction of paper and intensity of ink and colours. Residual acidity is detrimental to the permanence of paper.

In, 1997, the effect of neutral sizing was studied in laboratory by Paul, S.K. and colleagues. They observed that neutral sizing on paper promotes its durability. Poly aluminium chloride (PAC) has emerged as the primary polynuclear chemical for neutral sizing. On hydrolysis PAC produces high molecular weight and a large number of positive electric charge which is able to interact with cellulose fibres and size, optimising sizing and retention of fibers.

Sodium silicate alone or in conjection with rosin and starch has gained popularity for its excellent results. The cementing action of the sizing material is worth mentioning. Its resistivity to oils makes it fit for sizing of printing papers and papers used for prosters which have to resist the coloured and oily inks. It keeps down fluff, improves rattleness and resist the deteriorating action of sunlight.

Starcth was the only known material to the paper manufacturers of the past as sizing agent. It improves physical strength of paper in one hand, and its permanence and transparency on the other.

The gelatin sized papers are much more durable. Largely used typing papers are mostly impregnated with solution of gelatin which improves characteristics of paper making it more ink, colour and water resistant, imparting durability. Increased degree of sizing, however, confers curling tendency in paper. The process of sizing, therefore, requires a great deal of care to be controlled strictly.

Coating paper with sizing materials is not a modern concept. It was prevalent in medieval period too. Alexandra Sotariou, mentions in her text that early Persian paper manufactured in Samarakand was coated with vegetable sizes obtained from wheat starch, gum trugacanth or the boiled bults of asphodels. The materials selected for sizing have been changing from place to place and region to region, perhaps much guided by their availability. The starch, produced not only from wheat and rice but also from corn, potato, sago, arrow root or plantaint was employed. Dard hunter saw the use of rice starch in Kashmir in contrast to the wheat which was common in Punjab.

Rajendra Lala Mitra, in his report on the manuscripts in India, mentions that the century paper sized with yellow arsenic and an emulsion of tamarind seeds was followed by rubbing with conch cells. Further, while mentioning Kashmiri paper, he reports that country paper being rough was thickly sized with tamarind seed emulsion to make paper liable for writing. The paper for ordinary writing was sized with rice gel. Further, he states, that paper sized with this material attracts damp and vermin of all kinds and pest silverfish which thrives on it luxuriently.

According to Mr. Mitra European paper was inferior to country paper so far, as durability is concerned. This paper was liable to be rapidly destroyed by insects, termites and climate. Several vols, written on foolscape paper dating from 1820 to 1830 housing in Asiatic society had decayed in olderd and touched in several places by silverfish.

The John letter papers of European origin, though thicker and stouter were so far injured. Ink had faded becoming illegible from many places. Contrary to this arsenicised paper of Mss. of college of Fort William even being much older than European paper are quite as fresh and new as it were initially. Some papers were dyed with turmeric nearby 1945 or before.

The sizing technique in handmade paper industry was applied in parts. First one side of the paper sheet was coated and dried and then the remaining side. A size prepared by boiling ground wheat floor in water was used for the purpose.

Use of glue or gelatine made from animal hides, bones,

horns and hooves was also prevalent in handmade paper industry in medieval India. Sizing was simply pplied over the paper surface with implements such as a wooden handled tool covered with cloth and an oval shaped chamois like cloth.

Glosiness on paper is extremely an old requirement. It was achieved by hand polishing for which a paper sheet to be glossed was continuously rubbed by hand with a glazing tool. The most common polishing implements were agate quarts of flint stone with exceptionally smooth surfaces large shells, cyproean tigris (Kawati), polished glass, the smoothered horn of stag and even smooth wooden cylinders. These were called gota (Ghota) or batta. Skilled polishers however, were known Mohurahdar and in Ahmedabad they were called Morkhus.

Colouring of paper, has been a widely accepted practice of the Indian paper makers under hand-made industry. Though coloured paper is now produced for use in texts, albums, calligraphy and art and festival decorations, etc., yet for silver and gold writing the coloured surface (blue & black) was a necessity. Gold and silver writing in the past had a rich custom. Kangyurs, Manuscripts, housed in Tawang Monastery, Arunachal Pradesh are the best known examples of local made coloured papers for gold and silver writings. Besides, a deep golden yellow coloured heritage (Hartad) paper is found in Nepal and some parts of India is also of this category. Soot from fire places and chimneys was employed for colouring paper. But coloured paper so produced was cheap paper (blotting paper). Materials of plant origin such as onion, violet indigo, myrtle, barks, fattron, blue lotus, oak apple, and pomegrannate with addition of lemon and vinegar have also been used for colouring paper.

Varieties of Paper and Properties

Y.K. Rao, in his book, *The paper industry in India, Status and Prospects,* has divided paper broadly on two categories i.e. cultural paper and industrial papers. However, the Directorate General Technical Department (DGBT) has classified paper in following five categories:

1. Printing Paper,
2. Writing Paper,

3. Packing & Wrapping Paper,
4. Other Papers, and
5. Paper boards.

Writing and printing paper are known as cultural papers. Writing papers require brightness, a smooth surface and the ability to absorb ink without showing any feathering tendency. Printing papers, however, require a high degree of opacity. The basic requirement of wrapping and packing papers is strength. But they are machine glazed (MG) to have one side a glossy surface permitting printing.

Other papers are technically called speciality papers, which are generally used for electrical insulation, wrapping cables. Besides, security paper, cigarette papers, filter paper and air mail papers are the fine examples, of papers of this category. The list of varieties of paper and paper board may be seen at appendix III.

Numerous are the properties of paper but few of them are very important from the stand point of preservation and restoration. The knowledge of these properties help a conservator to select a right paper suiting to the requirements of preservation and restoration. Smooth surface liable for fantastic writing and painting was the only property of paper known to the people of olden times. But to day, it is known for various properties, i.e, physical, chemical, optical, electrical and microscopical. The important elements of physical properties are;

Weight

Thickness, density and stiffness are characterised by physical properties of paper. Weight of paper is expressed as gram per unit area instead gram per unit volume. It is known as G.S.M. (Gram Square Metre).

Strength of paper is conferred by the length of fibres, purity of pulp and nature of loading and filling materials. However, it responses differently in machine as well as in cross direction which are very prominent in machine made paper.

Direction of Paper

The direction of paper parallel to the direction of flow of the pulp stuff while the paper is under making is called machine direction. Because of flow of sieve, the various fibres of pulp try to set themselves parallel to the direction of flow of machine and, therefore, the strength of paper is always greater across machine direction. The direction perpendicular to the machine direction is cross direction.

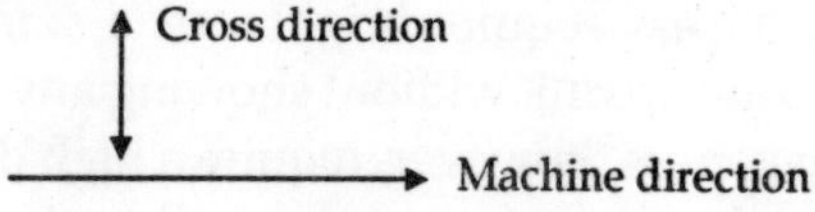

Figure

Sides of a Paper

A paper sheet is consisted of two sides, namely, wire side and top side. The side of the paper sheet which remains in contact with the sieve during the process of making is termed as wire side while opposite side of this is known as top side. The wire side contains mesh like structure which is examined under oblique illumination using low power lenses. If the structure is not visible clearly, the moistening of side with water or dilute caustic soda solution makes the examination easier as the moistening causes fibres to swell and making mesh structure clearly visible.

Water Marks

Incorporating marks in paper of animal origin or geometrical or anything else is very old practice of paper manufacturing industries. With what other purposes, these marks are incorporated for but authenticity and quality assurance of the product are the two objectives behind it. Words like *Bond Paper, Ledger Paper* are very famous water marks being used in Indian paper today. During examining the records proceeding vols, housed in N.A.I. repository, it was observed that the paper of records belonging to the years 1928 are water marked by two type of geometrical symbols and characters indicated below. These symbols are considered today important source for dating records. These marks are incorporated in paper with suitable device at the time of paper manufacturing process. Problem is faced to identify the right paper mark because duplicate papers with impressed water mark are also being sold. However, the original water mark is identified using ether. This chemical when applied on the mark, if duplicate, disappears while original remains unaffected.

Durability and Permanence

Durability of paper is defined as its ability to withstand handling while permanence the capacity of paper to withstand time keeping qualities. The paper which is tough and rattle does not mean that it will last without losing its strength for a long period of time. For example, newspaper when it is new, pronounces high degree of strength but by storage months together, it suffers discolouration and reduction in strength. Contrary to it the few grades of modern machine made paper such as "bond paper" normally retains its qualities for years together without showing any sign of discolouration and embrittlement. Hand-made paper used in 17th century on prior to that, still after centuries, seems young untyred and rattle. No doubt, such papers of some of the manuscripts have tired and turned brown specially in the written margins and the reason behind is the high acidity of inks used for writing and not the ingredients of paper. Hand-made paper of today also confers strength, retains qualities even in rigorous storage conditions. Obviously, two types of behaviour is pronounced by paper, one belonging to the category of newsprint and another by handmade or bond paper. The paper pronouncing character of first category is durable while that of second one is permanent. Durability of paper is considered a physical characteristic while permanence the chemical one. It is said that a paper of permanent value may be durable but a durable paper may not be permanent. Paper made from all rag pronounces a longer life than that made of coloured rags and plant fibers. Paper doubly treated with rosin sizing is as stable as that treated with animal sizing states Gunther Reichord.

A permanent paper, however, should conform the following physical and chemical requirements.

(i) Physical requirements:

(a) Folding endurance under one kg. tension	-	Not less than 250 double folds.
(b) Bursting Strength	-	Not less than 25

(ii) Chemical requirements:

(a) pH value	-	Not less than 6.5
(b) Rag contents	-	All rag. 100%

(c) Alpha-cellulose contents - More than 85%
(d) Copper number - 2 maximum
(e) Ash content - 2% maximum
(f) Rosin content - Not more than 1.5%

Sometimes, the permanence of a paper is ascertained by measuring folding endurance of unaged and aged paper sample under one kg. tension. If percentage retention of folding endurance of aged paper under one kg. tension is more than 70%, it may be deduced that paper has permanent character. The simplest formula for evaluating percentage is:

$$\text{Percentage Retention} = \frac{\text{Folding endurance of aged paper}}{\text{Folding endurance of unaged paper}} \times 100$$

The inherent strength of paper depends upon the length of cellulose fibres and their freedom from impurities like lignin which form organic acids with passage of time in suitable climatic conditions and make paper acidic.

Copper Number

The copper number of a paper is expressed as the number of milligrams of metallic copper which is reduced from cupric hydoxide to cuprous oxide in alkaline medium by a given weight of 100 gm. of cellulose material. The highest copper number a paper may have is 300.

Manufacture of Newsprint, Cheap paper and Boards in India

Newspapers, today one of the most important and influential vehicles of world countries, are supposed to reach the masses and hence requires that the paper on which they are printed should be exceedingly cheap. The marvellous development in the art of printing today have contributed very largely to this requirement and this has been supplemented by the production of cheap paper for the purpose, known as newsprint.

The production of cheap newsprint could be possible in India using cheap mechanical wood pulp, commonly also known as ground wood pulp. As the term signifies, this pulp is obtained by the mechanical process of grinding wood. No chemicals are used in its production except in very few and limited cases for bringing up the shade of the pulp. Mechanical pulp fibres, therefore, contain all the constituents found in raw wood. They become very short because of grinding and abrassive action of a grinding stone. Mechanical pulp, is, therefore, much poorer in strength and durability than chemical pulp. The mechanical pulp, is much inferior to chemical pulp and comparatively much cheaper in cost. It is partly because in its production, the cost of chemicals, steam, etc. is entirely eliminated and partly because percentage yield of wood (90%) is nearly double that of chemical pulp. Because of cheapness, Mechanical pulp has found a very extensive use in the manufacture of paper required for ephemeral purpose for which strength, quality or durability are not the essential requisites. Examples of such products are newsprint, cheap printing papers and cheap quality card-boards, etc.

Until 1936, there was no newsprint manufacturing facility in India. The reasons behind were many but the first was the total consumption of the papers in India was too small to justify the newsprint mill in India. Secondly there was lack of adequate suitable raw materials in British Indian Forests for production of mechanical pulp at economic prices in localities where the manufacture of this paper could be undertaken. However, the rapid economic and political development of the country and the spread of literacy among the masses began to foreshadow a vast and growing scope for the indigenous production of newsprint and cheap "mechanical" paper for writing and printing purposes. Owing to these facts, an experimental grinder equipment, to carry out investigation on the production of mechanical paper was, therefore, obtained and erected at the Forest Research Institute, Dehradun in February 1937. According to decisions taken by the Second Forest Utilisation Conference held at Dehradun in March 1937, the project for testing of various species apart from bamboo was recommended and approved.

The objective behind investigation was to test a number of species and select a few the most suitable, for manufacture of mechanical pulp so that the silvicultural possibilities of raising plantations for sustained supply of adequate quantities to a newsprint mill might be considered by the various provincial forest departments. Unfortunately, out of 22[+1] test species, none excepting that of pieca morinda and Abies nindrop were found in any part of our country.

The experiments and tests were carried out in two stages. The first stage experiments were aimed at to fixing the optimum conditions of grinding to obtain the maximum yield of pulp with minimum percentage of flour or fines and possessing the requisite characteristics of fibre length, freeness, felting power and colour strength etc. The second stage of experiments were however, conducted to produce pulp under determined conditions.

The sample sheets of paper were made from the 14 species of woods and bamboos by the Forest Research Institute, Dehradun. Subsequently, they were tested and found that the look through, finish and strength of the samples were not upto the mark.

The paper produced from Fir (Abies Pindrow) and spruce (Picca morinda) yielded the best quality of newsprint paper. The strength of these papers was superior to that of an imported newsprint paper.

One issue of a local weekly paper was printed on the paper produced from the mixture of 70% mechanical pulp from Kydia calycina and 30% chemical pulp made from bamboo. The press as well as public considered and accepted the paper. Papers produced from mechanical pulp of woods; Pinus longifolia (Chir), Broussanetia Papyrifoera (Paper mulberry), Kydai calycina (Pula) and Excaecaria agallocha (Gengwa) were fairly light in shade. The interior strength properties of papers produced from mechanical pulp using broad leaved species of wood were primarily and mainly due to the short length of fibres. The length of fibres from coniferous woods, like fir and spruce are comparatively much longer.

Newsprint from Blue Gum (Eucalyptus Globulus Labill) and Wattle Wood (Acacia Decurrents Wild) in India

Coniferous woods which are mainly used in foreign countries for the manufacture of ground wood pulps for producing Newsprint were not available in India. Investigation, therefore, was taken up by the Forest Research Institute, Dehradun, to produce Newsprint from the woods of the Broad Leaved Species such as blue gum and wattle wood. Blue gum billets (about 3' in length, 7" to 10" in diameter) approximately of 15 years old trees and wattle wood of 9 years old were received from Nilgris Division, Madras State.

Under the variables of burr, sharpness of stone surface, pressure of the wood on the stone and pit temperature, pulp was prepared from the wood trees aforementioned and standard sheets were made from 100% ground wood pulp and from a mixture of ground wood pulp (70%) and bamboo sulphate pulp (bleached) 30%. After conditioning at 65% R.H. and 21°C strength properties of the sheets were determined. The brightness of ground wood pulp varied from 44 to 53. The soaking of wooden blocks in water for varying periods and the retainment of pulp after grinding, before moulding into sheet were the principal reasons responsible for this variation.

The newsprint was also prepared from a mixture of pulps made from wattle wood, ground pulp and bamboo mixing in the ratio 3.1 and treating the mixture with 5% chlorine. See plates 2.6, 2.7, 2.8 and 2.9 for indigenous machines.

Old Manuscripts/Records

Among the old collections on paper support, one constitutes the 7,000 rolls recovered from a closed room of 476 caves in a hill in Central Asia. These rolls were discovered by Francis Franks and are presently being housed in British Museum, London. The significance of the collection is that they still seem young while they bear the date of 406 A.D. Besides, four Sanskrit Granthas belonging to 5th century A.D. discovered by Baber from a place Kugiar located 60 miles away in the south of Yarkand city in Central Asia constitute another old collections on paper support. Likewise, Chinese Manuscripts translated from Sanskrit, relating to Buddhist canonical literature on paper, at present housing in British Museum are also considered among the oldest collections. They are in roll forms, and hardly seems affected with the age though they belong to period 406 to 995 A.D. These were also discovered by Francis Franks.

The paper is very sensitive to the ambient conditions. The details about pronouncing effect on the material, with the passage of time by various elements is available in Chapter 3.

However, it changes its colour, hardens, curls and cockles as time goes by.

The Writing Fluid Ink

Ink is the another very important component in making of documents. Term ink, even after being such a homely and popular word, has not its origin in English Language. It originates from old French "Enque" to the Latin encastum, the purple red ink with which the Roman Emperors used to sign their edicts. The general concept for a layman about ink is a colouring fluid, which acts as a link between support and written characters. No research work has so far been done by any scholar revealing the facts of origin of ink, basic ingredients of early ink and spelling out the name of its discoverer. Harold S. Burt, in his article "Inks and Papers for use in Government offices in U.S.A.", has stated that the man has been efforting to find a perfect ink which would adequately convey to posterity an account of his actions, hopes and feelings. Further he mentions that according to carvalho, the first writing appeared between 1500 and 1800 B.C. However, Prajapati, in his book, "Archivo-library Materials, their Enemies and Need of first Phase Conservation", has mentioned that a Roman Physician, Aurelias Concelius Genyus and other contemporaries used "Atramenti" to heel cuts and burns as well as for the purpose of writing in 25 B.C.

So far ingredients and process of preparing ink, the suspension of pigment or colouring agent in appropriate concentration in water would have produced the ordinary ink. The possibility of using the process for preparing early ink by primitive men cannot be ruled out, but no sources to corroborate the fact are available. On the basis of experience, it may be said that the early ink was the suspension of two ingredients referred to above. Subsequently, as per requirement, the addition of binding agents which facilitated adhesion and fluidity to ink was started. Summing up the whole gamut, the pigment or colouring agents, medium and binding agents came up as the basic ingredients of a suspension technically called an ink early or modern.

About early inks, the statement of Harold S. Burst, "The earliest inks were those made from the juices and berries; then more ambitious souls gathered shells from the seashore which when grounded fine and mixed with water or the liquefying agent became the conveyor of the scriber's message to a limited posterity", conveys a very meaningful message and confirms the idea stated above.

Use of pigments by Indian seems to be very old i.e. more than 17000 years. The various types of colours such as red, ochre and green were used over rock paintings in India. In Mesolithic period, white hue joined the group of colours. The oldest rock painting of Bhimbaithaka, in Modern Madhya Pradesh, about 60 kms. away from Bhopal are known for their attractive colours and human figures as the subject for paintings of an artist. Even after centuries, the fastness of colours of these paintings is still bright and attractive. These paintings have been dated 17,000 years old. Obviously, the people of the period of Bhimbaithakan society were aware about the use of pigments, and binding material. It is the role of binding agents only which keeps the paints still adhered firmly to the surface. No doubt, the paint in composition differs a bit from the ink. But more or less basic ingredients of both are same. Making these facts the basis, it may be deduced that discovery of paint and ink was the same root of origin, but according to requirement of artist and writer, the suitable changes in density, fludity of the two accrued with advancement of time by adding more pigment and solvent.

R. Lala Mitra, has referred the inks of two kind. One fit for paper and another for Palm leaves. About the ingredients of first ink, Mr. Mitra states, it was lamp black, coffee coloured infusion of moistened rice, sugar and juice of plant called Hesurte (verbesind scandeus). The process of preparing ink is combursom and lengthy as it required continuous trituration in a mortar for several days before mixing the lamp black amulgamated thoroughly with rice infusion. For the want of sufficient trituration, sedimentation of lamp black leaving the infusion the top had occurred, hence making the ink unfit for writing. To add gloss to the ink, sometimes acacia gum was mixed with the ink. But this practice was not common as the sugar was performing the role played by acacia as well. Subsequently, slight changes in the ink formula through addition of infused emblic myrobalan prepared in an iron pot, was made. During the process of preparing infusion, two acids, gallic and tannic were added which formed salts of iron gallate and tannate adversely affecting the paper surface on which ink was used writing for.

The ink for writing over palm leaves was, however, composed of the juice of the Kesurte mixed with a de-coction of alta. It was highly exteemed as it sank into the substance of the leaf making it difficult to wash off.

Mr. Mitra has also mentioned that both the inks were of lasting character having no adverse effect on the substance of paper and leaf being free from mineral contents and strong acids. They never faded and retained their gloss for centuries.

The ink, however, used so far may be classified in two groups:

1. Writing ink, and
2. Printing ink.

The writing ink may further be sub-divided in following sub groups:

1. Carbonacious ink,
2. Iron Gall Ink, and
3. Ink with Provisional Colours.

Carbonacious Ink

Carbonacious Inks are the suspension of colouring agents, binding agents and solvents. Lamp black, soot or charcoal have been used as colouring agents while wine, water and vinegar as a solvent and gum arabic as binding agent in early days. In subsequent periods, glue and gelatin were also employed for the purpose of binding. Binding agents emulsified pigments in the solvent, provided viscosity keeping carbon particles in suspension and acted as binder to hold the practicles of pigements to the support on which writing was made.

India ink or China ink which are also known as manuscript ink, were prepared from pulverised charcoal. Out of three charcoals, (animal bone, sugar and wood) the wood charcoal which is 90% carbon was employed for preparing ink. The incomplete burnt substance of organic materials, having black colour is known as charcoal. The most commonly used ink in ancient Indian time was "Masi" says M.C. Joshi in his article, "Ink, Paper and Art of Writing in Perspective of Indian Antiquity." The word is derived from Sanskrit word "Mas" means pounding or crushing. This word, frequently occurs in the Granyasutras belonging to 200 B.C.

The inks of this category were prepared by beating lamp black or soot or charcoal with binding agents (gum Arabic or glue or gelatin) in a mortar until become pliable. Subsequently, moulded into required shape; tablet or stick. To prepare ink suitable fraction of these was, mixed with water or any suitable solvent in proportion (not fixed) at the occasion of writing.

The most common black carbon or cabonacious inks are of two varieties, ordinary or delible and indelible. Previous was prepared from pulverised charcoal mixed with solvent (preferably water) and gums for temporary writing. The permanent or indelible inks however, were deployed for writing of permanent nature. It was prepared from strong boiling lac mixed with water. The other ingredients of the ink were borax, lodhra (Atree) and lamp black or soots of sesamum oil. For writing over birch-bark another type of ink was prepared by mixing and boiling charcoal from almonds and cow-urine.

Recently, during visit to Majuli, the largest river island of the world, resting in the lap of river Brahmaputra, in Modern Assam, while examing the riches of Manuscripts, on Sanschipat housing in different satras of island a new type of local ink used for writing over the bark was seen. As for, the formula for preparing the ink, has not been brought to light by any scholar so far. According to Secretary of the committee, Shri Kanak Datta, looking after the welfare of Garmursatra, the ingredients of the ink are Haritri (known as *Harad chhoti* in Hindi language), a herb 'Ghamira' (in Bundeli language) and urine of cow. For preparing ink, Haritri and ghamira were powdered separately after drying in shadow. Then the two were mixed in cow urine and boiled. However, Shri Datta failed to supply the information about the proportion of the constituents of the ink. The work on the preparation of ink in laboratory and to study its adverse effect on the plant fibre, if any, is in hand. It is expected that the results of the same would be published soon.

The gold and the silver inks have also been prevalent in Indian societies. Tangyurs, now-a-days, housing in the Monastery of Tawang, are the best known examples of Manuscripts written with these inks. Not only in Medieval Period, but in ancient too, such inks were deployed by scholars for artistic writings, rich patrons or sacred books in India as well as in Tibetan areas. Besides, ancient texts of Jain Scholars pertaining to religion and astrology (Almanacs and horoscopes) have made constant use of variety of coloured inks mostly of herbal origin. There is need for a research on the variety of inks used in the past in Indian subcontinent and neighbouring region.

There is a great deal of controversy among European scholars specially from Britain and U.S.A. about the Indian ink. Few believe that no ink was invented in India. But this view of scholars is somewhat imbalanced, and biased. The use of colours, origin of art of writing in Indian sub-continent are rich evidential sources to corborate the truth that India is the only country claimant of world's oldest and the richest cultural heritage. Mitchell, a well known authority dealing with ink studies has nowhere disputed the origin of Indian inks. He has used the term Indian ink or Chinese ink for the inks which were prepared from lamp black.

Inks of this category neither fade with passage of time nor adversely affect the support like paper. However, two capital imperfections of the inks are washability and smudging in dampness.

Sepia ink is another writing fluid belonging to the inks of this category. It is prepared from a black brown pigment discharged by cuttle fish. She discharges this pigment in order to cover retreat when attacked or terrified. Ink prepared from this pigment got the name sepia after the name of her mother fish. The permanence of this ink is questionable i.e. after a period of 50 years or so it fades.

The two major imperfections, washability and smudging, paved the way for the discovery of indeligible ink or permanent inks. One such ink was prepared from lac mixed with water, horax, lodhra plant (tree with white flowers) and lamp soot of sesamum oil boilding them strongly. Perhaps, this ink has been not widely used to the reasons unknown. However, the widely used ink of indeligible character was iron gall ink. The first ink of this category was prepared from all the ingredients of carbonacious ink and ferrous sulphate. Addition of latter provided the ink a character of permanence. Too much quantity of ferrous sulphate in ink however, resulted in making written characters brownish which was not regarded by the scholarly world. The unwanted change of colour was corrected by addition of galls++ as ink ingredients.

Inks with Provisional Colours

Blue, blue iron and red inks fall into the category of this type of inks.

The discovery of blue inks, actually wrote a new chapter in the history of inks. First time Indigo was deployed for making of the ink. According to some European Scientists, Indigo was first made in India from the leaf of Indigofera Tinctoria, a member of the order Leguminosate. As Mitchell believes, indigo was first used in 1770 for the purpose of preparing ink. Water and gum arabic were the other ingredients of this ink besides indigo. However, no such formulas, telling about the ratio of different constituents of the inks has been seen so far.

Dr. A.K. Gauda (Poona) tried to find the date of first blue ink but he did not come out with any specific results.

Smudging in dampness and washability in water are the chief demerits of the blue ink. However, colour holding capacity, non-damagial character to the paper fibres and remaining unaffected by certain micro-organisms are regarded as the principal merits of the ink. Because of these merits, the sphere of users continuously increased with the passage of time, inspite of noted demerits and even after discovery of blue iron inks. It is believed that indigo prohibits the growth of micro-organisms. The hue of the ink shows remarkable behaviour during machine lamination. It turns bright blue which gradually retains its original colour after few weeks. People are divided in their opinion about the causes for change of colour. Few believes that it occurs due to the interaction of small fraction of acetic acid which remains as residual in the acetate film and ingredients of ink. The acetic acid changes the blue

Indigo into Indigo white in presence of high temperature. Others believe that the sulfurous acid present in the acetate foil is responsible for this change of colour. Whatever be the actual cause of change of the colour, it is not permanent.

Blue Iron Inks

The discovery of ferricoferricyanide in the last quarter of 18th century paved the way for manufacturing of Blue Iron Inks. The compound is marketed with trade name Prussian Blue.

Finally powdered Prussian blue when compounded in suitable proportion of gum and water, gives the blue iron ink. The ink in character and behaviour differs from Iron gall inks. High ratio of iron in Iron gall inks is detrimental to paper while no such evidence has been exhibited by the blue iron inks. These are resistant to bright light and bleaching agents. However, they are adversely affected by moderately strong alkalies and smudge in dampness. Many of the writing with this ink are marked by high degree of lustre even after lapse of 100 years or so.

Coloured Inks

The use of coloured inks was prevalent in olden times for writing rubrics, colophones, important words and for marking ends of the chapters. Some coloured fluids were also used for correcting errors. The ink made of cinnabar or alta was used for the first purpose while for latter a colour made out of red or yellow orpiment, grounded in gum water was employed. The quotations from texts were smeared over with a little ochre in commentarial works in general in every rigion but in North-Western provinces in particular. A thick black varnish was used for writing over Palm leaves.

Red Inks

The use of red ink was prevalent in Indian society in 4th century A.D. It is evident from the narration of great Sanskrit Poet Kalidash, in his famous text Kumarsambhava. He has compared a Bhojpatra written with red ink with motifs of an elephant. To colour the trunk of an elephant with "Geru" (a red colour soil) at special and religious occasions was a custom prevalent in Indian society.

Extract of Brazil wood has been used widely inceptionally for preparing ink of red colour. A formula for preparing ink of this colour, states that first of all one ounce of rasping of Brazil wood, and two grams of each white lead and alum are pulverised and fused in one pound of urine (cow perhaps). When this material is added with a drum or two scruples of gum arabic gives red ink. Another formula, suggests that the respings of Brazil wood are left standing either in vinegar or urine (cow perhaps) for several days. Subsequently, when a little amount of gum arabic is added that gives red ink.

Studies of William F. Begget have thrown light on the fact that Brazil wood was known to the people of Medieval period and the same was imported from India.

Another raw material other then Brazil wood deployed for producing red ink was "Vermilion" which is chemically known as mercuric sulfide. Though it is insoluble in water but when fine powder of the compound is added in appropriate quantity of water or vinegar, pronounces red colour. To add to viscosity and binding property to the ink, suitable amount of gum arabic or glare of eggs is mixed with it. This type of red ink was very much in use in Rome.

The significant character of this type of ink is that it does not smudge in dampness. The brightness of hue of ink, however, shows dullness turning reddish brown with the passage of time. The fresh writing of the ink can easily be washed in water but attaining the age of two hundred years, it adopts indeligible character almost. Obviously the permanency of ink increases with the time.

Modern Indian Inks

To cater to the requirement of users of today, various inks of brilliant colours are being produced. Most of them got popularity because their stains over clothes and hands are easily removed. Such inks were formulated and manufactured because of the increasing demand of fountain pens. But experience of their impermanency and difficulty of deciphering written character over yellowed paper, compelled ink manufacturers to formulate inks of permanent nature. Inks of this category are known as "Ferrogallotannate" fountain pen inks. Two variety of inks of this category are:

Ferro-Gallo Tannate Fountain pen ink with,

(i) 0.1% iron content, and

(ii) 0.2% iron content.

The constituent materials to compose blue back ink with 0.1% iron content are Tannic acid (4.0g), Ferrous sulphate crystals (5.5g), Gallic acid (1.5g), concentrated hydrochloric acid (5.0g), phenol (1.0 ml), Dye ink blue (5.0g) and distilled water to make up the total vol. to one litre. For preparing inks other than blue black, provisional dye, in place of dye (ink-blue) is added.

So far as procedure for preparing ink of this category, suitable quantities of Tannic and gallic acids are added in 300 ml. of distilled water and heated to 60°C in a beaker until they dissolve. Solution so prepared is filtered in one litre flask and marked I. Likewise, ferrous sulphate crystals are dissolved in a separate beaker in minimum amount of water and concentrated hydrochloric acid is added. Solution of ferrous sulphate is added to the solution kept in flask marked I. Now phenol solution in separate beaker is prepared in minimum water and added to the flask containing, mixture of solutions. So prepared mixture is cooled to room temperature and dye solution is added to it. To mix up the solutions homogeneously flask is shaken and distilled water is added to bring the total volume to one litre. Solution is allowed to stand over night then filtered through a Buchner filter. Now ink is ready.

The constituents of ink with 0.2% iron contents are the same as that of previous ink but their ratio is different as shown below. It is manufactured using the same technique as stated above.

(1) Tannic acid-450 g.

(2) Ferrous sulphate-10 g. crystals.

(3) Gallic acid-1.0 g.

(4) Concentrated 6.00 g. hydrochloric acid.

(5) Phenol-1.0 g.

(6) Dye-ink blue 5.0 g.

(7) Distilled water-Remainder to make up the total volume to one litre."

The part played by each constituent of ink is different and significant. The Dye imparts hue to the ink, whereas hydrochloric acid keeps the constituents of it suspended preventing sedimentation. Ferrous sulphate, Tannic acid and Gallic acid form a ferrogallotannate compound that is black in colour. Phenol, works as preservative and hinders the growth of fungus and bacteria. Water serving as medium provides fluidity to ink constituents as a whole.

Keeping qualities of both the varieties of ink are marvellous as they can be used well with a fountain pen without showing any clogging effect for a minimum period of two years from the date of manufacturing. However, it is safer to store them in amber coloured bottles. They are known for showing permanent character against ultra violet light.

Other Modern Inks

Now-a-days various types of inks such as liquid ball point pen and fibre-tip inks, felt-tip inks, ball point pen pastes, dry writing ink, typewriter ribbon ink and stamping ink are prevalent which have their role in creation of archival documents.

Liquid Ball Point Pen and Filter Tip Pen Inks

Such inks normally are the aqueous solutions of anionic dyes. To prevent the writing tip from drying out 20-30% highly boiling point solvents such as ethylene glycol is added with the inks of this category.

Felt Tip Pen Inks

In general, two types of felt tip pen inks are available. The constituent elements of one type of ink are dyes (anionic and cationic) and alcohol while that of another type are soluble dyes in aromatic hydrocarbons. Resins, soluble in alcohol or toluene are used as binding agent in both types of ink.

Ball Point Pen Pastes

These are 40% solutions of dyes (including basic dyes) in fatty acids like oleic acid or acidic resins (Phthalate resins). Sometimes, neutral solvent soluble dyes are also used for preparing such inks. Benzyl alcohol or glycerol, low volatile products, are solvents and polyvinyl pyrolidine (resins) are thickening agents in inks of this category.

Most of the inks or pastes are water resistant but they are affected in acetone and alcohol.

Dry Writing Ink

Copying pencils and coloured leads are covered under this category of inks. They are mostly composed of a dry lead incorporated in a wooden stick. Graphite for ordinary pencils, organic pigments for coloured pencils and dyes (cationic or anionic) for coloured copying pencils are used. Pencil lead contains ceramic as binding agent. Natural as well as synthetic polymers (water soluble and insoluble) are used for coloured and copying pencils.

Typewriter Ribbon Inks

In 1900 A.D. when typewriter was invented ink prepared from carbon black and fat was used for ribbon. The ribbons were imprignated with the mixture of two. But today, development has brought about many changes in ribbon technology and major two types of ribbons i.e. woven and foil are very popular. Woven ribbons are imprignated with the mixture prepared from carbon black or pigments or basic dyes and non-drying oils whereas dye is used in transferable plastic or wax.

Stamping Inks

Two type of stamping inks; oil free and oil suspended are being manufactured today. Previous type of ink is composed of water/ethylene glycol and dyes (anionic or cationic dyes) while latter is prepared from pigments or fat soluble dyes in oils and fatty acids.

Effect of Inks on Paper

Carbon inks have no adverse effect on paper. However, irongall inks have been found detrimental to paper vehicle. Charring of paper on written margins and yellowing are the result of adverse, effect brought out by iron gall inks. It is not only observed in manuscripts but in printed books also. Recently, while examining few printed old rare books housing in the library of Indian Institute of Advance Study, Rastrapati Bhavan, Shimla, I found few printed sheets darkened in the written margin and bitterly damaged. The paper from few characters was eaten through. The sheets of some volumes of Military

Department Index belong to first quarter of 19th century, at present housing in 6th floor of Annexe of National Archives have suffered because of change of colour, charring and the ink impression has passed through the paper.

The question is why iron gall inks are detrimental to records. The simplest reply to the question is the acidic nature of these inks. Since there was no definite proportion fixed for mixing constituents of ink for their preparation, people applied their own formula and prepared ink. Lure of fast colouring, mostly divulged people in adding larger quantity of ferrous sulphate which in result, imparted more acidity to ink.

Acidity works as a slow burning agent. Higher is the acidity faster and deeper is the reaction. When inks of higher concentration were used to write upon, they brought about charring and yellowing of paper sharply. However, inks with low acidic contents were not sharper in reaction and this is the reason that few manuscripts and printed documents are still safe after centuries. The elite class of scholars perhaps seems aware of the fact and that is why they, without any attraction of colour, preferred ink with low percentage of ferrous sulphate which pronounced low acidity and brought less damage. Studies in this respect, reveals the fact that Lewis who first wrote about high acidity in iron gall inks concluded that the major cause of paper deterioration is high acidity present in it, which was imparted by iron gull inks.

Behaviour of Inks with Time

The studies on inks reveal the fact that some inks change their colour with passage of time. The change of colour is governed by various factors, such as the constituent materials of ink and heaviness of hand of writer, the kind of pen point used for writing and conditions in which document is stored. But normally oxidation of constituent materials of ink in atmospheric oxygen and time span are the major causes of it. For instance the fresh writing with a blue ink would always appear blue but after a few hours of writing, it turns dark. Further, exposing this writing to diffused light for few weeks, the colour of ink appears blue black. The behaviour of blue ink was tested with varieties of paper of different quality in the same atmospheric conditions but the result always confirmed the above. It is obvious from the fact that no other factor but oxidation of ink constituent is responsible for change of colour. However, exception to this was writing with the ink on paper coated with alkaline clay. Writing on such paper soon after writing turned black. The reason behind is that the acidity of ink is neutralised by alkali present in paper and colour appears black.

Ferrogallotannate dye based inks are also affected with the oxidation with the passage of time showing a colour change. Inks of this category follow almost a set pattern. First they turn blue-black which subsequently changes into brownish black, dark brown and finally fainting to pale brown. In the process of disintegration, ferrous changes into ferric sulphate then to ferric oxide while dye completely get oxidised and disappears. But due to residual ferric oxide, rusty lines on the written characters are left.

The change of colour of Ink with the time is technically termed as ageing of ink which is affected by atmospheric conditions specially light and storage environment. Longer exposure of documents in light and storage in dampness accelerates the process of ageing because of which the process of colour change completes in short period of time than normal course.

Determination of Age of Ink

Determination of age of ink is basically a subject of forensic science to detect forgery. But it is equally important in an Archives to prove authenticity of documents sometimes. The question is whether or not the date of ink can be determined exactly? It is a problem which in general can not be answered by one factor alone. However, two principal techniques, one making of chemicals and taking colour factor into account, have been tried by the conservators of world to answer the problem. The application of both the techniques, require careful handling of documents under age determination. For details, see, at page 165 Dating.

Revival/Deciphering of Faded Writing

Revival/deciphering of faded writing is a popular practice in the forensic as well archival world today. For the purpose, both dry and wet methods are used. Exposure of faded documents in Ultra Violet light and infra-red light facilitates deciphering the faded contents. For convenience, the photograph of such documents is prepared for permanent retention and use. Santucis technique, which is a wet method, however, has been founded very effective in revival of faded writing. In this technique document is subjected in various solutions, such as water, 2% ammonium sulphite, saturated solution of basic lead acetate, 1% solution of acetic acid and finally so treated document is washed in clear water for 20 minutes. It is believed that the solutions used in the technique, have no adverse effect on the document.

All the types of faded writings neither can be deciphered nor revived. Only those inks which contain iron give positive results to some extent.

Erased pencil writing, nigertine and Chinese ink (Carbon) can sometimes be deciphered by exposing the document to iodine fumes.

Colours: Composition of colours, slightly differs from inks. For depiction of ideas colours of various kinds have been deployed in painting and illustrated manuscripts. See Annexure I.

REFERENCES

Alexandra Soteriou, *Gift of Conquerors*, Hand Paper Making in India.

Basu, Purendu, "Enemy of Records", *The Indian Archives*, Vol. IV, No. 1, 1950, pp. 7-13.

Bhandari, K.S. and Srivastava Anita, *Wet Cleaning of Straws for Improving Pulp Quality*, "IPPTA", Vol. 10, No. 3, July-Sept., 1998, p. 159.

Buhler G. and others, "Indian Palaeography, p. 26.

Cultural Heritage of India, Vol. III, Ramakrishan Mission, 1953.

Dutt, Hirendra Nath, *The 20th Century Cyclopedia of Ink and Allied Products*, Vol. I, p. 4.

Experiences, the Researcher Gained during his 25 years of Service as Conservation Scientist in *National Archives of India*.

Gunther Richardt, "The Durability of Paper", *The Indian Archives*, Vol. 1, No. 3, 1947, p. 231.

Harold S. Burnt, "Inks and Papers for use in Government Offices in U.S.A.", *The Indian Archives*, Vol. X, No. 2., 1956, pp. 13-16.

Indian Forest Bulletin, No. 217, Cellulose and Paper 1957, pp. 1-13.

IS 1774-1986 (Refirmed 1990), *Standard Paper for Permanent Records*.

IS: 1581-1975, *Ink for Permanent Records.*

IS: 220-1988, *Ink for Permanent Records.*

Joshi, M.C. & Deys, "Ink, Paper and Art of Writing in Perspective of Indian Antiquity", *International Seminar on Examination of Documents,* held in Madras, 1997.

Kathpalia, Y.P., *Conservation and Restoration of Archive Materials,* p. 23.

Kirk, RTF, *Paper Making in the Bombay Presidency,* A Monograph.

Lal, B.B., *Scientific Examination of Documents, Works of Arts, Monuments and Other Cultural Property,* Conservation of Cultural Property in India, pp. 24-25.

Mahajan, Sushma, *Upgradation of Wheat Straw and Rice Straw for Pulp and Paper Making,* "IPPTA", Vol. 10, No. 3, July-September, 1998, p. 93.

Mahalingam, T.V., *South Indian Palaeography,* p. 103 "Nearcus one of the generals of Alexander during his invasion of India said that people of Punjab knew the art of manufacturing paper out of cotton and tottered cloth."

Mitra, Rajendra Lala, "A report on the manuscripts in Indian Repositories", *The Indian Archives,* Vol. 2, 1948, pp. 27-28.

Mitra, R. Lala, "The Report on the Mss in Indian Repositories}, *The Indian Archives,* Vol. II, 1948, pp. 21-32.

Paul, S.K. and Coleagues, *Neutral Sizing of Bagasse Furnish, A Laboratory, Study,* "Indian Pulp and Paper Technical Association" (IPPTA), Vol. 3, Sept., 1997, p. 61.

Prajapati, C.L., *Archivo Library Materials, Their Enemies and Head of First Phase Conservation.*

Prajapati, C.L., *Archivo-Library Materials, Their Enemies & Need of First Phase Conservation,* pp. 53-74.

Prajapati, C.L., "Conservation of Mss. in Tawang Monastery", *The Indian Archives,* Vol. IXXIX, No. 1, 1990, pp. 25-28.

Pralekhon, "Pandulipion Tatha Dusprapya Pustakon Ka Abhirakshan Aur Pratisonkar", *National Archives of India,* 1992.

Prasad, Rajendra, *High Grade Paper for Records and Other Printed Materials of Archival Interest, Conservation of Cultural Property of India,* Vol. IX, 1976, p. 34.

Rajesh, K.S. and Rao Mohan N.R. Bagasse, *The Promising Alternative for the Future,* Indian Pulp and Paper Technical Association, (IPPTA), Vol. 10, No. 3, July-Sept., 1998, pp. 151-57.

Rao, A.R.K., Bhattacharya, J.P. and Banerjee, Saugata, *Current Status of Indian Paper Industry and Outlook for Investment in Near Future,* "Indian Pulp and Paper Technical Association" Vol. 10, No. 4, Oct-Dec., 1998.

Rao, Y.K., "The Paper Industry in India—Stattus and Prospects".

Satyendra, *Pandulipi Mala,* p. 91.

Sharma Chhaya, Rao, N.J. and Singh, S.P., *Effect of Recycling on Fibre Characteristics,* Indian Pulp and Paper Technical Association, (IPPTA), Convention Issue, December, 1998, pp. 69-73.

Talwar V.V., *Loading and Filling,* Conservation of Traditional Records Materials, p. 68.

Utilisation (new Series) Indian Forest Bulletin, No. 108, 1941, pp. 1-10.

+1 (1) Excaccagia agallocha
(2) Macarauga Spp.
(3) Broussonetia papyritera
(4) Bombax malabaricum
(5) Erythrina suberosa
(6) Buteu trondosa
(7) Boswellai Serrata
(8) Eucalyptus globulas
(9) Sterculia spp.
(10) Abies Pindrow
(11) Picea morinda
(12) Kydia calycina
(13) Gmelina arborea
(14) Pinus longifolia
(15) Tetrameles nudiflora
(16) Mallotus Phillipinensis
(17) Teak Saplings
(18) Trewia nudiflora
(19) Stereospermum suaveolins
(20) Acacia decurrens
(21) Garuga Pinnata and
(22) Odina Wodier

++ "*Galls* are the dwelling places of a type of insect called gallwasp. It appears as tumour wen gallwasp punctures the bark of oak tree for laying eggs. Galls are rich in tannic and gallic acids which are stored by the adult gallwasps for food to their insect babies"

3

SCIENTIFIC STUDY OF THE FACTORS PRONOUNCING CAUSE OF DETERIORATION AND DAMAGE OF ARCHIVO-LIBRARY WEALTH

The importance of records and why they merit expenditure of time and money for their preservation has been discussed in previous chapters. The question of Preservation of the Wealth, however, when figures, it flashes in the mind as to what is to be done under the subject. The simplest answer to the question is to ensure physical defence of the wealth against the elements identified as the destroyer of the wealth. Such elements, in the field of conservation are termed as the enemies of records. The next question arises as to what are they? To find out the perceptible reply, it would be better to quote Mr. Purvendu Basu who has mentioned in his article, "Enemies of Records" that the principal enemies of the wealth are time, fire, water, light dust, heat, humidity, atmospheric gases, fungi, vermin and last but not the least human being, K.K. Guha Roy, in his article, "Library of the Indian Agricultural Research Institute" has mentioned that industrial smoke, acidic pollution of atmosphere, dust, heat, light, moisture and design of the building apart from insect and fungi are main factors which quicken the pace of destruction.

The adverse effect pronounced by the elements, however,

on the documentary wealth is not similar in nature. Some bring about embrittlement, discolouration, hardening, reduction in flexibility, fainting of ink, hue and colour while some erose surface and make scars in the form of holes, tunnels, stains and tears. The deleterious effect of some on record is slow invisible and remains unidentified till they are subjected for tests, manual or mechanical.

For systematic study of the factors pronouncing causes of deterioration and damage of wealth, they may be classified in two groups which are sub-grouped as shown below:

(1) The Enemies responsible for decay of support materials.
(2) The Enemies responsible for causing physical damage to the vehicles of information.

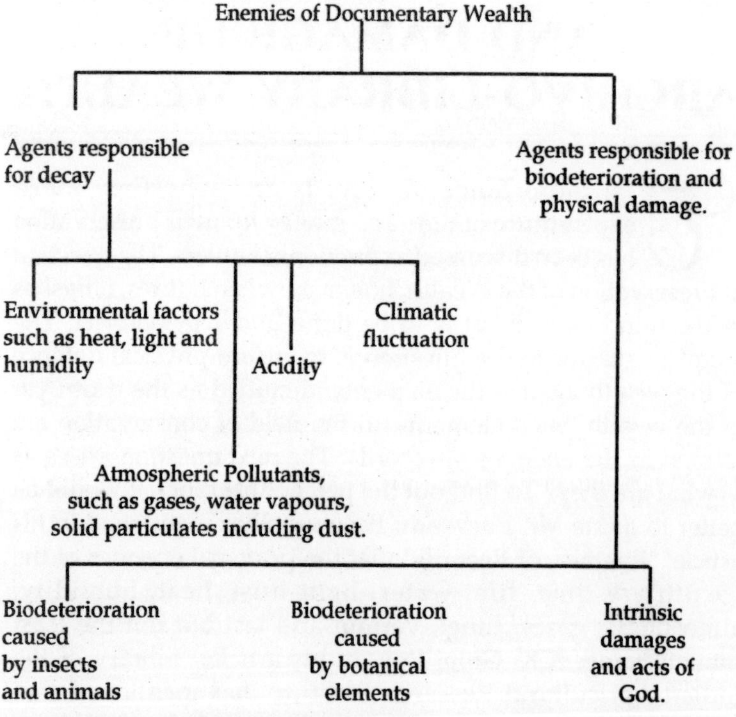

Chart 3.1 : Enemies of Documentary Wealth

Agents Responsible for Decay

Decay is the process in which, vehicle of information loses its strength with the passage of time without showing any scar on the surface. Sometimes, however, colour of few grades of paper changes due to decay.

Light, heat, water vapour, atmospheric gases, dust and acidity are the prime agents causing decay. Nothing is wrong to term these elements as silent killers of record materials.

Light

Sun is the bounting source of natural light. Besides, this energy is harnessed through artificial sources, electricity is the most prominent of them. Natural light is associated with various types of radiations, visible and invisible among them are ultra-violet radiations; infra-red radiations and light radiations helping visibility. The U.V. Radiations vary from 136 A° to 4000 Å in wave length and form "far" and "near" two regions. The radiations between 4000 Å to 8000 Å from visible radiation are responsible for producing day light. The radiations, however, in the range between 8000 Å to 4×10^6 Å are known as infra-red radiations. This region is termed as Heat-Ray Region also. All these radiations associated in light, natural or artificial are electromagnetic in nature. However, some artificial lights such as glow light are free from ultra-violet radiations while fluorescent light does not.

Though all types of radiations including ultra violet radiations, visible radiations and infra-red radiations, more or less, are destructive to paper and other type of organic fibers, yet former one is known for its destructive potence as the radiations being of short wave-length carry heavy energy. As such they not only penetrate easily but also are absorbed by most of the substances readily causing luminescence and disbalancing internal set-up in turn setting up photo-chemical reaction in cellulosic material. Once the reaction starts, that continues round the clock even in darks. The paper made of ground wood pulp gets easily affected with the radiations and undergoes both the embrittlement and discoloration while paper composed of 100% rag fibres, 100% sulphite pulp and the mixture of the two though susceptible to deterioration but without any visible evidence of discoloration.

O.P. Aggarwal, in his article, "Environment and Preservation of Paper" has mentioned that both visible and U.V.

radiations cause damage to paper although at the higher wave lengths the activity of light is less.

The natural light is rich source of U.V. radiations but among the artificial lights, the light produced by fluorescence and bright light are associated with these radiations. Light produced by the tube lights is the best example of this kind.

Paper sized with resins, containing impurities of copper and iron, discolour readily with the effect of light. Rosin is destroyed by oxidation under the influence of light, says Gunther Reichardt in his article "The Durability of Paper".

Discussing the chemistry of deterioration of paper and cellulose material by the effect of light, the glycositic bonds between the monomeric molecules forming cellulose polymer are broken. Because of it, cellulose compounds of lower molecular weight are formed as cross-linkage develops which results in disturbing the original fabric of paper, and brings about discolouration, embrittlement, loss of flexibility, hardening of material and loss of strength. Some grades of paper however, do not discolour due to decay. Obviously, discolouration in paper, which is sometimes presumed a serious warning for conservation scientists from the view point of preservation of records, is not always true for all category of records. Discolouration of paper occurs not only due to adverse chemical change in materials in making of paper but also due to other factors such as presence of either or all impurities like chromophor, iron and copper in paper and also because of accumulation of dust and smoke particles amalgamated with tarry materials greases and storage in damped conditions.

Launer and Wilson conducted a practical study on the effect of light over paper and revealed the fact that the light usually in the range of 330-440 millimicrons affects paper adversely.

The impact of light over the paper and other organic materials depends on its intensity, distance between the source of light and object and the time of exposure. A Committee appointed by the International Council of Museums for studying the problem of Protection of Museum objects from light, pointed out that apart from above the impact of light also depends on the external factors (heat, humidity and gases) present in the ambient atmosphere, the spectral characteristics of radiations and the ability of mateials to absorb light.

Scorching Sun bakes the pages rendering them brittle. For convincing proof about deleterious effect of the sun need not to go to a scientist or laboratory. If a sheet of paper under test is torn into two and one piece is kept in shade and the other exposed to the sun, the change in colour and tensile strength of the two pieces will be apparent even to a layman observer after a few days.

Heat

Heat and its variation have deleterious effect in making paper brittle which crumbles into dust after some time, says Purendu Basu, in his article, "Enemies of Records". Further, he states that it results in expanding and contracting cellulose fibres over and over again, thus weakening them. Today's study on the impact of heat over the cellulose materials has gone a long way, but observations of Mr. Basu being of preliminary nature, based on indigenous source and belonging to the early period (1950 A.D.) when the preservation in Indian sub-continent was setting a foob and taking a shape, are regardable.

The impact of temperature on paper and other materials of organic nature, not only causes their degradation but also sustain fungus and insect growth, drying of adhesives, yellowing of paper, buckling of films and tapes, loss of flexibility and hygroscopicity because of dehydration. Heat has more degradable influence on cellulose and other materials in making paper brittle than acids and oxidising agents. Heating of cellulose to elevated temperature say about 200°C for a long period of time, brings about a loss in paper structure. Greater is the heat intensity severe is the effect. Barrow's study of heating paper to 120°C, 100°C, 80°C and 60°C make him to observe that the time taken to reduce the folding endurance of paper for a given value is 7.5 time longer for each drop of 20°C.

Richard Smith studying the effect of temperature on the permanence of paper has observed that after each half life the original fold value of paper is reduced to half. One folding endurance half life has been defined as one year at 60°C (140°F) for the purpose of estimating half life at lower temperatures.

The values of half life estimated by Smith are given in the table below :

Table 3.1: Values of Half Life of paper estimated by Smith.				
Average Storage Temperature		*Paper Half Life (in years)*	*Average Storage Temperature*	
°F	*Diff from 68°F (in °F)*		°C	*Diff. from 20°C (in °C)*
140	+72	1	60	+40
122	+54	4.1	50	+30
104	+36	18	40	+20
95	+27	40	35	+15
86	+18	88	30	+10
77	+9	204	25	+5
72.5	+4.5	320	22.5	+2.5
68	—	490	20	—
63.5	-4.5	760	17.5	-2.5
59	-9	1200	15	-5
50	-18	3100	10	-10
41	-27	7900	5	-15
32	-36	21000	0	-20
14	-54	170000	-10	-30
-4	-72	1700000	-20	-40

Smith in his studies about the inter relationship of temperature and acidity on the properties of paper worked out paper half-life with varying values of PH and temperature. He showed that both temperature and PH have great effect in deterioration of paper.

O.P. Aggarwal is of the opinion that the effect of heat on paper also depends, on other things, such as the nature of fibre in making into paper, acidity and other contaminants present in the paper. It is also observed that the effect of heat on paper is not equally attributed to all grades of paper. Papers containing impurities of iron and copper are readily yellowed by the heat. Further, ground wood pulp paper is more susceptible to heat than good quality paper made of chemical pulp.

Obviously, the heat is considered serious problem for archivo-library materials in tropical and sub-tropical regions.

Water Vapours

Water vapours present in the environment causes humidity. Paper and other allied materials being hygroscopic in nature, absorb water vapours from the surrounding in close proximity, which results in swelling of paper tissues, softening of adhesives, staining of paper and growth of micro-organisms. The more absorption of vapours in damped areas, sometimes results in bleeding of ink, limping loss of rigidness and sticking-up of paper sheets forming a solid mass. O.P. Aggarwal, in his article, Environment and Preservation of Records, had mentioned that too much vapours not only act as physical agent but also as chemical agent accelerating most chemical reactions. The natural ageing of paper in humid conditions is much more faster than otherwise. Fungi and mildew grow rapidly in high humid conditions.

The higher humidity is an essential adjunct to the process of bio-deterioration. Lower humidity is detrimental to paper and other organic materials as it results in deccication of paper. Documents and objects like paintings (miniature), wooden objects, bone and ivory are weakened due to fluctuations of humidity. It also causes internal stress and strain in the fibrous material affecting their strength. Too little humidity tends to dry cellulose fibers and rob them of their resiliency while too much humidity encourages the mould growth.

The tropical and sub-tropical countries where climate is marked by too high and too low humidity during rainy and summer seasons respectively, are victims of moisture generic problems in so far as documentary wealth is concerned. In Indian sub-continent alone, some parts (north, north-east and coastal areas) reel under high humidity during major part of the year in general, and in rainy season particular. In some areas such as western and central parts including Madhya Pradesh, Uttar Pradesh, Rajasthan, etc. temperature rise is marked by 45°C and above during summer and low humidity touching 15% - 20%.

Joint Effect of Heat, Light and Humidity

Launer and Wilson in their studies found that humidity or water vapour influence the degradation of paper in presence of sun light. They, however, noted that this effect was true to the paper made of rags, cotton cellulose and had a reverse effect on wood pulp paper. Studying the effect of drying of wet paper in fire, sun and normal conditions (shadow) it has been found that yellowing and hardness developed over the wet paper dried in fire is greater than the dried in the sun and shadow. The paper dried in normal atmospheric conditions is the least affected.

Atmospheric Pollutants

Atmosphere is represented by various gases, dust particles of various origin, smoke and moisture. These representatives of atmosphere, differ quantitatively and qualitatively from place to place and region to region. For instance, the concentration of polluting gases, smoke and solid particulates is found always higher in the atmosphere of metropolitan and industrial towns and cities. The obvious reason is that in metropolitan cities, various types of vehicles, means of transport and industrial units using petroleum and coal as fuel produce gases and smoke. Similarly, fine dust particles are present in every type of atmosphere prevailing in tropical and sub-tropical countries but particulates of geological and chemical nature assimilates with the atmosphere of the regions where factories vibrate, volcanoes burst and quarrying functions.

Gas Pollutants

Though atmosphere is represented by various gases, yet gases responsible for polluting atmosphere, and hampering record materials, are oxides, nitrogen, sulphur dioxide, carbon dioxide, carbon monoxide, oxygen, ammonia, ozone, hydrogen sulphide and peroxides. These gases are also known as noxious gases.

Approximately 78% nitrogen is present in the atmosphere. Nitrogen at elevated temperature produces its oxides because of lightening, heating of furnaces and boilers. Besides, in Archives and libraries, Cellulose nitrate base films are identified the rich source of producing nitrogen and its oxides. Obviously, in the areas where industries are very thick, the atmosphere of such areas will be rich in nitrogen oxides. Out of five oxides of nitrogen, nitrogen oxide and nitrogen dioxide are responsible for damaging papers.

Combustion of coal, coke, pyrites and oils are rich sources of producing sulphur-dioxide. These are widely used in industries and means of transport for running them. Carbon mono-oxide and carbon dioxide are also produced by combustion of various types of fuels. Purendu Basu, in his article "Enemies of Records" has mentioned that sulphur dioxide even in a dilution of 0.5 to 1.0 part per million part of air is readily absorbed by paper fibres.

Higher humidity in atmosphere, paves the way for access of various types of oxides i.e. oxides of nitrogen, sulphur dioxide and carbon dioxide to record material as these are readily soluble in water. These oxides mixing with the water present in atmosphere form their respective acids and are absorbed by the paper because of its hygroscopic character. Oxides of nitrogen, sulphur dioxide and carbon dioxide when dissolved in water form. Thus, the oxides in question, impart acidity to paper and other allied materials and acidity them. Nitrogen dioxide also bleach dyed leather and fabrics. It has been observed that the edges of pages of old books and volumes are more acidic than the interior portions.

Ammonia being soluble in water, generally gets converted into Ammonium Hydroxide which can potentially affect various kinds of inks adversely. Now-a-days, because of rapid industrialisation, this gas is formed as by product of various chemicals and increase its concentration in the atmosphere. Its presence can be experienced in the areas with poor sanitation. Ammonia is harmful to cellulose (alkali hydrolysis). Free ammonia reacts with other chemicals in presence of water and forms Ammonium Sulphate which deposits on the surfaces of materials varnished with natural regions giving rise to a "Bloom".

Hydrogen Sulphide is a very effective gas for disfiguring images over microforms and paintings. Bio-degradation of proteins containing sulphur is the main source of producing hydrogen sulphide. It is also produced as a by-product of many chemical plants.

Hydrogen Peroxide being a powerful oxidising agent, induces internal change in the fibrous material like paper, textile and destibilize them. Degradation of drying oils which are getting place in Archives and Libraries in many ways and processing of exposed photofilms, are identified the main sources of producing peroxide in Archives and Libraries. Photographic films and prints are vulnerable to hydrogen sulphide, hydrogen peroxide and sulphur dioxide.

Ozone is an unstable gas, but it is identified as the potential destroyer of documentary wealth. It is formed with the interaction of oxygen in presence of electric sparks and discharges of ultra-violet light. Ozone layer depletion because of fluro-chloro carbons and photocopying machines are some sources of producing this gas. It destroys organic materials by breaking the bonds between carbon atoms. Paper, textile and other fibrous materials on long exposure to it, lose their strength. It oxidizes varnishes which have been used by early book binders to give high gloss to leather covers. Gelatin, paste and glue, particularly in humid elimate are degraded by ozone. Natural rubber is extremely valuable to it.

Nitrogen oxides, ozone, sulphur dioxide and hydrogen sulfide bleach photographic images and subsequently attack gelatin present in the emulsion and the filmbase.

Dust Particles, Smoke and Aerosole

Free circulation of air in the repositories specially located in tropics, causes considerable quantity of dust to accumulate and that is further augmented by the installation of central heating. Dust particles consisting of silicon and soil of geological origin generally represent the atmosphere of tropical and sub-tropical regions. Other particles consisted of cement, lime and minerals also join the atmosphere in the areas where volcanoes bursts, factories vibrate and quarrying functions. Besides, smoke particles generally produced because of incomplete combustion of fuels like coal, coke, oil, wood and other rubbish materials make their way to the atmosphere. Though smoke particles of black colour are essentially carbon and chemically inert, yet they being surrounded by oily and tarry materials may be quite acidic. These suspended particles enter the building and accumulate over all surfaces including books, volumes, files, racks etc. and the edges of these articles, exposed to ambient environment. The tiny, hard and with sharpened edges, the particles of silica, embedded over documentary wealth, rub against the surface while handling or even when is a draught of air causing abrasion in the paper fibres.

Solid particles of dust and gases in presence of water vapour, in the atmosphere collide because of dispersion. Particulate matters such formed are known as aerosols. The size of such particles varying from 0.2 to 0.5 millimicron, seems very small to be seen by unaided eye or ordinary microscope. Mist is yet another form of particulate matter of liquid dispersed in gas phase and called liquid aerosols.

According to Professor Morris, Particulate matter of both categories are chemically mixture of smoke or tarry organic material, inorganic matter and coarse dust. Anthracene, Aromatic and carcinogenic hydrocarbons of the pyrene, benzene fluoranthene and phenanthrene belong to the group of smoke and tarry organic matters while metal compounds of alkali, carbonates, chlorides, fluorides, nitrates, sulphates: light and heavy metal oxides of inorganic nature belong to inorganic matter. Likewise, proteins and other organic materials of high molecular weight, hetrocylic compounds, oily and solid organic materials are covered in the group of organic matter. Alumina, clay, fly-ash, silica gypsum and calcite, etc. are the elements of latter group.

The particles having radius less than 0.5 millimicrons, follow Stoke's law of flowing fluids of spherical nature. As such, these particles do not fall on the ground by their weight but sweep randomly this way or that by every current of air. Such particles are carried by gushing winds and storms. Air molecules with other suspended particles bombard the wall, surface and any thing coming on their way damaging them. Sometime, they stick to their surface. Smoke particles are very quick to stick to the surface of cloth, paper, leaves, because of electrostatic attraction and if they are not cleaned timely such particles settle over the surface forming layer over the layer. Water vapour help to make them stick on the surface which result in discolouration of paper, affecting legibility of letters and painted images adversely.

Walls of temples, Gompas and other religious monuments where ritual of Havana is a daily feature, generally become dirty. Cloth, books and paintings housed in such vicinities also discolour. It happens because of accumulation of smoke particles over the surfaces of respective objects.

According to survey conducted by Mr. P.K. Yennawar, Mr. S.N. Dixit, on the pollution status of four cities of India, it was reported that Delhi contains 700µg/m(3) suspended matter in air. It was found to be more not only compared to Indian major cities (Mumbai, Kolkata, Kanpur), but also with world's major cities like Chicago, London, Cincinnati, San Fransisco and Tokyo. Surprisingly, the second polluted city that was Mumbai followed by Kolkata, with suspended matter consistence of 527 µg/m(3), is also from India. Though the survey in question was conducted 26 years ago in 1977, yet so far as pollution status of above cities is concerned, there is no any soothing improvement as impression given by day to day reports publishing in Newspapers and Magazines. So far as Kolkata, the leaves of trees turn black during winter because of accumulation of smoke and other matter particles suspended in air. In some parts of Delhi, near I.T.O and old Delhi people start vomiting because of high pollution. In such a prevailing situation of pollution, the future of archivo-library materials in such cities in particular and in the areas striked by high pollution contents in general, is highly questioned.

Mr. S.P. Singh, in his studies on "Particulate Matter in Air and Works of Art" has mentioned that organic materials like textiles, leather, wood are most susceptible to surface action by particulate matter. Here it can be added whether it is paper or parchment, palm leaf or vellum, they are equally affected by the agents in question.

Soluble salts from sea water form another source of solid particles dispersed in air alongwith water particles. The solid inorganic salts as colloidal particles deposit over all kinds of objects, including documentary wealth and wall painting located in coastal areas and adversely affect them in the long run in many ways.

Acidity

Acidity is the major threat for the life of machine made paper. The few pulping techniques being used by modern paper mills make use of acid cooking liquor (sulphite cooking liquor). Apart from this, some filling and loading materials being used by the mills to bring about desired surface characteristic impart acidity to paper. Such materials are calcium sulphate and zinc sulphate, etc. Sudborough and Mehta of the Indian Institute of Science, Bangalore on their study on Preservation of Cultural and Library Wealth in the libraries of this country, revealed the fact that acidity in paper is the one of the major causes of their deterioration. This acidity in some grades of paper is inherent because of the use of acid cooking liquors, filling and loading materials of acidic nature and sizing material like alum. Moreover, some other sources have also been identified which contribute acidity to documents. Such sources are use of acidic inks, treatment of paper in the arsenical materials and corrosive sublimate like mercuric chloride which have been used once upon a time as coating material for preventing books and documents against insect damage. Besides, wrapping materials like cloth, paper and file covers etc. having high acidic contents also constitute a potential source contributing acidity to documents, the reason being that acidity has transmigratory character. Acidic polluting gases present in atmosphere, are no exception to it. They also impart acidity to documentary wealth of all kinds. Lignin, in paper, yet cause another source of acidity.

Due to chemical action of acids, cellulose degrades into hydrocellulose, the compounds of lower molecular weight. Sulphuric acid of 78% strength gelatinises cellulose while concentrated nitric acid induces it in forming esters. Action of acids if continues for a long period of time, it result in transforming cellulose into friable white or cream coloured powder.

John Davis, in his article, "Preventive Conservation of Archieves had reported that over 75%, perhaps 90% of paper deterioration in library book collections caused by acid hydrolysis. Further, he had mentioned that the damage of documents and allied materials caused by acids is particularly serious in industrial areas where the atmosphere is polluted with noxious gases. In paper making process alum is extensively used for dispersion of the fibres in water and also to precipitate rosin size on the fibres. But too much use of alum results in enhancing the acidity range in paper. High degree of acid, present in paper leads to the formation of hydrocellulose, a product of lower molecular weight and consequently to the imbrittlement of paper and allied products.

Role of Environment

Environment is defined as the surroundings of the society, human beings, colonies or objects. In case of Archives and library material, environment means the atmosphere in which the documentary material is housed. The term denotes the factors like climate (humidity and temperature), light, solid particulates and microbiological agents. All these elements, which act upon the material constantly day in and day out, have a significant bearing on their preservation.

It is recognised that the environmental conditions, in which cultural objects like paper, leather, textiles painting are placed, greatly influence their deterioration. O.P. Agrawal has stated that the works of arts which were deposited during the World War II in repositories with controlled atmospheric conditions survived much longer and showed less sings of deterioration than those which were left outside. In general organic materials are much more susceptible to deterioration by adverse environment than the inorganic ones.

Role of Climate in Deterioration of Records

K.K. Guha Roy, Librarian, I.A.R.I. Library, in his article, Library of the Indian Agricultural Research Institute, has mentioned that climate and storage conditions have been amply demonstrated during the last ten years of the Library's existence in Delhi in 1936. The stack area was open to the afternoon sun, the west dry wind and dust storms, etc. which reduced the strength of paper to become brittle.

Sudborough and Mehta who carried out investigation on the perishing of paper in Indian Libraries during 1920-21 (Journal of the Indian Institute of Sciences, 119-226) have mentioned in their report that the paper has become quite brittle. Even books which have not been much used show the phenomenon and in many cases the brittleness is so marked that one or two single folds of the paper cause it to break along the fold. They further mention that perishing of paper is more common in India than Europe or America. While mentioning so they have quoted the fact that Mr. Chapman, Librarian, Imperial Library, Kolkata, in 1915 sent 8 volumes from Indian Libraries to British Museum, London. The copies of the volumes lying in Indian Libraries showed marked sign of perishment while volumes at British Museum were devoid of it. In one case the sign of discolouration and foxing was also observed.

This fact led Mr. Chapman to request professor Sudburough of the Indian Institute of Science, Bangalore, through the Government of India to undertake research into the problem of storage of paper, books and documents. A systematic survey was carried out and the findings of that were published in the journal of Imperial Library, Kolkata. Out of 140 Books examined, 100 seemed perished loosing 71% strength. Comparative figures of loss in other localities were, Shimla 9% Oatacammund 21%, Chennai 71%, Mumbai 68% and Meerut 63%. The reason behind the dis-similar degree of loss at different places was the climatic variations in different places. The climatic conditions of hill stations such as Shimla and Oatacamund is much more conducive for the preservation of paper than the climatic conditions in Chennai, Mumbai, Kolkata and Meerut.

Against this backdrop, a further survey was also conducted to determine the extent of damage caused to paper documents stored in stack areas in closed steel almirah/shelves. It is evident from the following tabular statement indicating the extent of loss that the old periodicals mentioned at Sr. No. 5 were the most affected part of the collections.

Table 3.2 : Extent of Loss Occurred in Paper Material Placed in Different Places of World

Sl. No.	*Name of collection*	*No. of volumes examined*	*No. of volumes perished*	*No. of volumes under way of perishing*
1.	Ann. Inst. Agron Paris (1876-1900)	16	4	6
2.	Phil. Trans. Roy Soc. (1871-1900)	30	3	15
3.	J. Chem. Soc., London (1862-1900)	72	14	17
4.	Amnalen der Chemil, Berlin (1832-1900)	162	7	26
5.	Agric. Ledger, Kolkata (1892-1912)	17	17	00

Discolouration is a feature indicative of the fact that process of deterioration has already started. But it is not true for all varieties of paper.

Starch and animal sixing is decomposed when exposed to damp and they are attacked by micro-organisms, says Gunther Reichardt.

Bio-Deterioration Caused by Insects and Animals

Deterioration caused by elements of zoological and botanical (Biological) nature is termed as bio-deterioration.

The seriousness of deterioration caused by insects can very well be manifested from the following statement of E.A. Buck, Bureau of Entomology and Plant Qurantine, U.S. department of Agriculture who, in his article "Book Worms" published in 'The Indian Archivist, Volume 1, No. 2, 1947 has stated.

"When one enters such thoroughly modern structures of stone, steel and cement as the Congressional Library of the National Archives of U.S.A. in Washington D.C. or the Huntington Library in California, to mention only three repositories of documents embodying the best information available to the Librarian, he is so overwhelmed by sense of beauty and permanency that he finds it hard to believe the often repeated statement that insects have destroyed more books and papers than fire and water."

Continuing his description on book destruction Mr. Buck further writes, "The book destruction by insects during the past 15 years, leaves little doubt as to the soundness of belief that insects are in the front rank of book enemies.

The role of insects as book destroyers is not a new episode of today but it had been recognised in the past even before Christ. Among the very early writers, Aristotle (+a) (335 B.C.), Horace (+b) (65 B.C.) and Ovid (+c) (43 B.C.) have mentioned about the creatures, damaging documentary material in their respective writings. Moses while addressing Joshua give instructions regarding the preservation of books of the Pentatench by anointing them with cedar oil and storing them in earthen vessels.

(a) "Aristotle writing in Athens in 335 B.C. mentions creatures in books, resembling grubs found in garments.

(b) Horace expressed the fear that his writings would eventually become "food" for vandal moths.

(c) Ovid, while in exile, likens the constant quarrying of sorrow at his heart to the Knawing of the Bookworms as the book when laid away is nibbed by the worms teeth."

The familiar, proverb, often commented to very studious person, "Book-Worm" was used by the Phillippus of Thessalonica, in first century A.D. in satirical comparison of grammarian with bookworms. For what purpose the World Bookwork was used by the author, is immaterial to us but it gives a very meaningful information that people of 1st Century A.D. were familiar with the insect "bookworm" which used to damage books. Ausonies Mokings, at the tutor who prefer to burry himself in worm eaten and outlandish scrolls rather than give himself more familiar pursuits and refers to a choice between preserving writings with cedarwood oil or all owing them to perish as food for worms, is the another example, manifesting the fact that people of Ausonius period (4th century A.D.) were aware about the damages caused by insects and their counter measures.

E.A. Buck, in his studies has classified the injurious insects and pests to the written and printed world in three groups; The true bookworms, Termites and surface feeders. Dr. S.M. Nair in his article "Biodeterioration" has mentioned that most common deteriogens of paper are Bacteria, Fungi and Insects.

Book Worm

Book-Worm or Gastrallus indicus, Reitter—a familiar name to Indians and Burmeize is found very active in infesting documentary material all over the world. Most specifically, it is also known as Sitodrepa Panicia belonging to the anobiidae family. The insect received its Latin name from its occurrence in dry bread (Panis) and in Europe it is known as bread beetle but its chief injuries are to druggist suppliers hence the name drug beetle. It is specially partial to pepper. In United States, it is known as drug store beetle, in France as Vrillete du Pain and in Germany as Broad Kaper. Anobium Peniccum is another popular name of book-worm. Swezy, Lasioderma Serricorne F, Catorama Herbarium Gosh (rarer), Ptilinus Testaceus Pic are some species of book worms, found in Indonesia, (Dutch East Indies) and Malaysia.

Houlbert, describing the various stages of book-worm as larva, chrysalis or pupal and beetle, says that they are extremely small probably about two millimetres in length. The larva has a cylindrical curved white body with dark mouth parts while beetle has a rather uniform brown colour. It is possible for drug store beetle to develop from an egg under favourable conditions, within the short time span of two months. Four generations of this species during a year are not unusual. Each female lays about 60 eggs at a time. Presuming half of the each generation is female, the pests multiply at the rate during first generation 30, during second generation 900, during third generation 27000, and during fourth generation 8,10,000. In other words, a female will multiply giving a large family of 8,10,000 members in a year. Really this multiplying rate of insect is alarming threat to a custodian of records or librarian. But, belief runs rough as a Houlbert believes that all the individuals do not reach full growth. Many dies in the larval stage by the fate of parasite or devoured by the carnivorous species while other disappear because of cold when they are not enough deeply embedded in the medium protecting them.

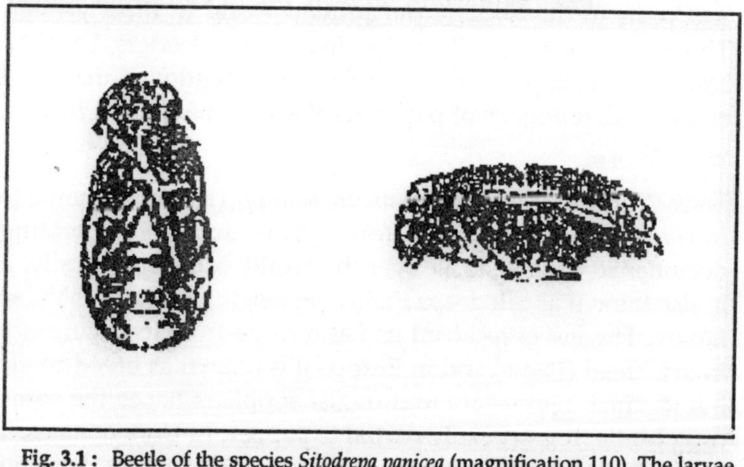

Fig. 3.1 : Beetle of the species *Sitodrepa panicea* (magnification 110). The larvae of this beetle are particularly found of old books and manuscripts. The *Sitodrepa panicea* belongs to the *anobidas* family. Its correct name is *Gastrallus indicus* familliar in India and Burma as the 'book-worm' and is responsible for the perforations and tunnels in the covers, books, documents and files of many libraries and record offices.

The adults which are brownish or blackish beetles are almost never seen without a close search. These are inconspicuous and seldom active in mid-day light. The certain adornments possessed by the beetles, make it possible to distinguish the species under microscopic examination. The white grubs (Fig. 3.1) or immature forms look alike to the average person. On disturbing by the turning of pages through which they have been building their tunnels, curl into light balls and roll out of the book or roll just enough from their tunnels to be crushed when the book is closed.

The grubs hatched from eggs, at once start burrowing into covers, preferring those in which glue, paste or casing is available. They frequently concentrate their attack along the back of hinges of the book covers, cutting the sewing threads used for keeping gathers together, thus causing the pages to fall apart. As grubs get more mature and voracious from these original points to attack, they extend their tunnels through covers and pages according to their habit. While doing so, they secrete a glue like material, lining tunnels and pupal chambers. This material later on, results in blocking the pages into a solid hindering separation. Sometimes, so blocked sheets require soaking in clear gasoline before opening.

In the beetle stage, it has wings to facilitate its movement from drug store to libraries. Eight-tenths of damage done in the bindings of the books may be attributed to bookworms. The female drug store beetle deposits its eggs near the surface of the binding on the edge of the leaves, choosing the most favourable spots for the evolution of their progeny. The full grown beetle emerges from its pupal casing at the end of 15 to 20 days and in order to free itself, they pierce with their jaws the film of silk which separates it from the outside. It takes wings, if temperature is favourable. The circular holes which are sometimes seen in great numbers on the back of old bound books and volumes are the exists.

When number of the grubs borrowing in leather bound books are left unattended for months without being removed from the shelving, they will emerge out from the holes made in the leather. The excrement of book worm/ grubs gathers in their tunnels and between the pages as a fine dust. It may vary in its colour according to the type of paper. This powder or dust will silt out when the badly damaged books are shaken over paper. The real destruction of all types is caused by dust making grubs.

About the nature of damages caused by the larva, Dr. Nair says "The larva activity was found to predominate within the book-bindings though this was not revealed by surface examination alone. The bindings mostly composed of raw, unbleached wood pulp often possessing thin layer of leather pasted on the surface are seen damaged by the insect. Number of round holes of 1 to 2 mm radius, and tunnels are made by the larvae. They also make parallel tunnels of about 1 to 1.5 mm in width running in all directions, during their feeding activity, see fig. 3.2. The tunnels or galleries are also found to be filled with irregular shaped foecal pallets. On the

Fig. 3.2 : Book worm infected document.

edges of the covers as well as in the pages, the concentration of galleries made by larvae is a normal feature." E.A. Buck has expressed that insect attack on the books and papers increases as the climate becomes warmer and more humid.

The adult beetles are small, slender and brown in colour. They measure about 2 to 3 mm in length and 1 mm in width. Interestingly, the mature larva has a stout, semicylindrical crescent shaped body measuring 3 to 4 mm with swollen thorax.

The gravity of disaster caused by the bookworms, can be

manifested from the two major problems due to which Royal Society of Cottingen in 1774 and the International Library in 1903 were compelled to offer prizes for a satisfactory solution.

It is revealing to many that sometimes books and manuscripts, badly damaged by insects, when examined, are not represented by even a single living bookworm. But there is no mystery, in this state of affair as nature's balancing act is worth commending. Nature has provided enemies of bookworms in the form of tiny parasites never seen by untrained eye. The remains of cacoons of parasites, in many a books free from bookworms but badly burrowed by them, tells a silent story of battle: supremacy between the bookworms and parasites occurred perhaps now or 100 years ago.

Bookworm infection of archivo-library material is reported from all part of the world but they are more active in subtropical regions. Though several authors, have described many other insects as bookworm, yet larvae of some beetles belonging to the family anobiidae are the true bookworms says Dr. S.M. Nair. Among the Indian states, Punjab, U.P. (Dehradun), Andhra Pradesh, Delhi, Gujarat, Kerala are the places where bookworm infestation is said to be very serious.

Booklice or Psocids

Winged and wingless psocids of several kinds are reported to exist world over. They are very frequently seen on books in some libraries and in many homes and are called lice because of their whitish tiny structure hardly as large as the head of an ordinary pin. Their importance as book pests has been exaggerated. The life cycle of psocids completes in 3-4 weeks at mild climate. Warmth and dampness favour their growth and multiplication.

Semi-transparent body having long structure like antenae and chewing mouth parts signifies the physical structure of the insect, mentions S.M. Nair. Further, he records that white or greyish white to brown coloured book-lice have length varying from 1.0 mm to 2.5 mm. Psocids are said to be frail creature incapable of causing physical injuries to book covers. They do not bite people, but carry diseases or harm books. They are objectionable only in the annoyance. They may cause nervousness to the persons who are not aware of the fact that they are harmless.

Booklice attack is more concentrated on leather objects than among the ordinary books. Szent Ivany, however, reports that Lipsoscelis a species of Psocids, has been observed eating gelatine of slides of a mounted Echinodermata. On the other hand A.W. Mckenny Hughes reports that book-lice or psocids feed on mould and mildew. As such, while feeding on the infected records by the mould and mildew, they remove the surface of paper or binding.

Booklice are almost omnivorous and capable of damaging books, paper and photographically devouring glue and starch as well as fungi on paper. It has come to light that natural silk, synthetic fabrics, leather and parchment are also damaged by the insect besides, cotton cloth, Cupboards, baseboards, near plumbing fixtures. The areas where dampness prevails are the places where it prefers to rest. The presence of booklice is a warning that the conditions are developing which may attract more dangerous vermins.

Cockroaches

Cockroach, a common household pest, is a surface feeder, seriously defacing book covers but rarely do more even when most abundant. The insect is fed on sizing materials such as starch and gelatine which are abundantly found on cloth, bindings and sized papers. E.A. Buck, in his article "Bookworms" has mentioned that the large American cockroaches may become very destructive in closed library spaces and may actually eat off the backs of cloth bound books or files. The insect emit an ink like liquid which further defaces books. A.W. Mckenny Hughes, records that cockroaches of different kinds usually attack the outside of books, making blottened marks or eating the gummed title from the back of leather or cloth. They do not often damage the pages. In tropical and subtropical areas or even in gulf coastal states, where the insect is found abundantly, the protection of documentary wealth is a continuing battle that is won only by eternal watchmenship and application of remedial measures.

Over one thousand species of the insect are existing in the world with majority living in tropical and sub-tropical regions.

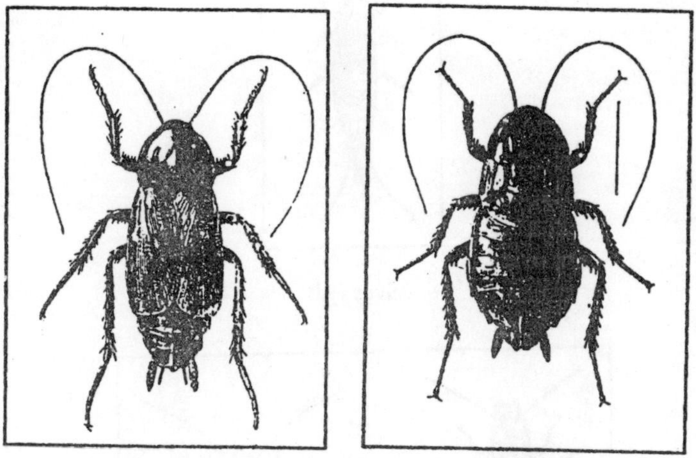

(a) Male (b) Female

Fig. 3.3 : Cockroaches

Fortunately, only few varieties are likely to be found in Archives, Libraries and other places housing records. They carry filth on their bodies and are suspected of spreading poliomyelitis, cholera, typhoid fever and dysentry.

Silverfish

Silverfish, a wingless, silvery coloured, carrot shaped insect of 8 to 10 mm, length resembling in its movement with fish. According to Mr. S.M. Nair, the body tapers very markedly from head to tail and is covered with thin scales giving rise to silvery shiny appearance. The young very much resembles the adults.

Temperature and humidity play a significant role in the development of this insect. The larval stage may last from three months to two years depending on the temperature and humidity. Humidity above 55% and temperature between 16 to 24°C, provide ideal conditions for the development of insect. It is identified as surface feeder having choice of starch, glue, gelatine (photographic) sizes and pastes as their food value. The backs of bound books are generally attacked by the insect as this portion offer them glue and starch. Illustrated manuscripts, wall papers, miniature, painting are also

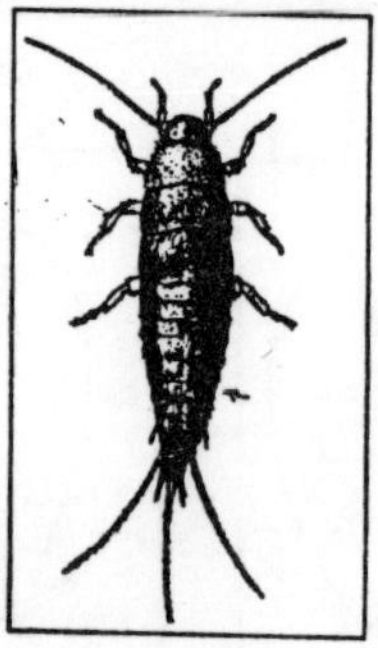

Silver Fish

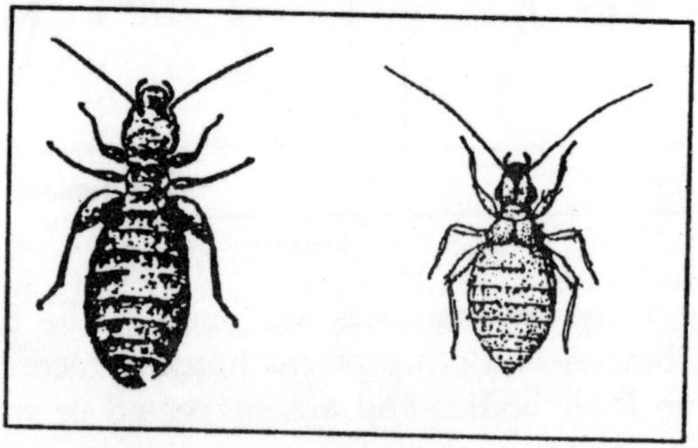

Book Louse

Fig. 3.4 : Silverfish

disfigured by the insect. Under favourable conditions, a female lays 10 to 50 eggs which are hatched within a time period of 6 to 10 days. Cockroach and silverfish are known as surface feeder.

Moths

Moth larvae are seemingly most potential damaging insects which may be found in the hinges of books and their back where glue, the material of their food value, is found. A.W. Mckenny Hughes has recorded that these, on occasion, make galleries in the paper penetrating many pages and frequently turning back to the binding board from which they started. Bound books, stacked close together, are generally attacked by the insect from outside. They may eat the surface in meandering strips, especially if the cover is treated with dressing containing animal fat.

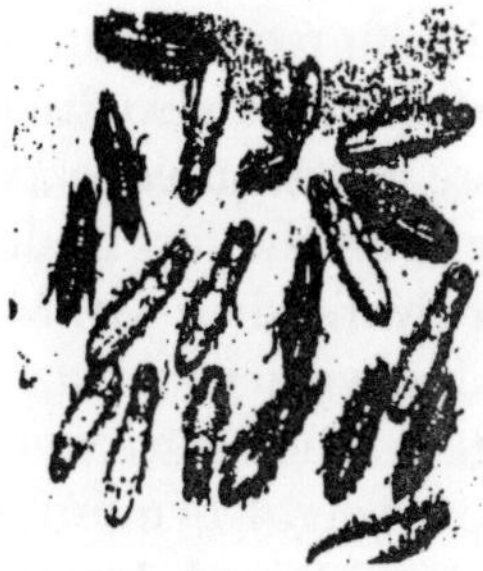

Termites

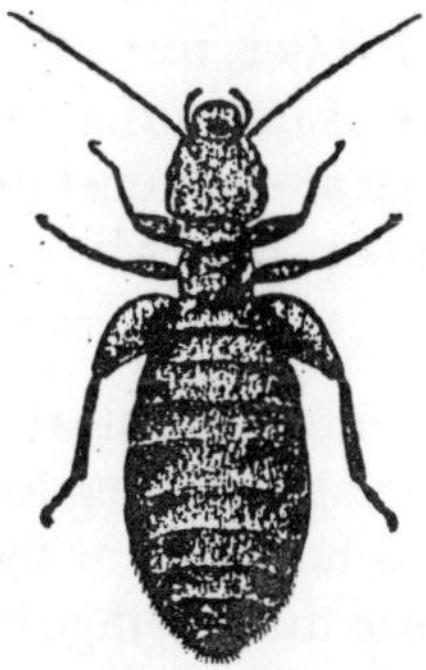

A Book louse (Greatly Enlarged)

Fig. 3.5 : Insects

The brown house moth is abundantly found every where in the world except South America and Africa. The larvae are hairless, dark-headed, white, 1.5 cm long worms. They eat cloth and sometimes leather covers of the books. The adults which bear wings, lay their eggs in cracks on the floor and wall and any other concealed places. In warmer climate four generations of the insect in a year are common.

Crickets

This insect is found very active in tropical and sub-tropical regions. A lot of food-grains during harvesting time of rabi crop is consumed by this insect. The attack of the insect is not confined to food-grains only but extended to pages of books and files also. The records kept in boxes, racks, not in frequent use are damaged by the insect through the process of knawing.

Mice

The mice are destructive to record materials as they not only knaw board, paper and cloth but also discharge a liquid which rapidly destroys texture of arsenised paper. Mr. Mitra's story about mice, throws light on the fact of destruction of records by the insect. A. Mukhtiyar, when Mr. Mitra was in his early age, put two fresh looking written and stamped sheets of paper in the bottom of the cage in which mice were captive. The papers were allowed to stand over night. In morning when paper sheets were taken out, they were stained, looking old and tired. Mr. Mitra relates his experiences further by mentioning that repeated washing of fresh paper sheets with a mixture of fluid discharged by mice and water, turns them in appearance old. The effect of treatment was not surfacial but perceptible to the interior of the paper. What are the constituents of the discharged fluid, has not been mentioned by Mitra.

Human Beings

Theft, vandalism and pilferage (stealing in piecemeal) are various acts of human beings bringing loss to the documentary wealth. Besides, careless handling at the various lavels right from creation of records to their storage in departmental record rooms and Archives, is a very common cause for destruction of records. In addition to these, bad appraisal, tampering, kleptomiacs and incendiarism also harms record materials by one way or the other. Every scrap of paper produced cannot be kept for ever in a creating agency as they all are not so valuable to merit expenditure of money and energy for their preservation. So at some stage selection has to be made to sort out the records of ephemeral nature so that they can be destroyed without any harm to the users. Purendu Basu, observes that bad appraisal often led to the valuable records being thrown away and the value-less kept. There are some people who may wish to tamper or distort the evidence.

Bio-deterioration by Fungi

Fungus infestation is a serious problem for preservation of archivo-library wealth comprising of paper. So far as the varieties of fungi, attacking paper material, are not less than 100. The important characteristic of fungi is that they have remarkable capacity to dissolve cellulose. Surface erosion, causing fluffiness, staining and weakening surface of material on which they grow are the obvious results of fungoid activity. Black, brown, greenish and even colourless (white) stains developed over the surface of paper, cloth, wood, and film indicate fungus attack. During inspection of books housed in the Library of Indian Institute of Advanced Studies Shimla, it has been observed that a variety of white fungus resembling white milk cream, has infested books, leather and wooden articles housed in damped CPWD room. Occurrence of white, black or a brownish coloured fungus is a common phenomenon in central, north and north-eastern zone of Indian sub-continent.

Chromatic alterations of record materials affected with fungus growth is really an interesting subject for research. Fawta Gallo's, studies, however, on the subject have mentioned that process of pulping, acidity, impurities, in paper (copper, iron, etc.) and length of fungoid action are responsible for it. According to Waksman, fungi attacking books and paper belong to the generae Aspergillus, Acrostalagmus, Cheetomium, Eidemella, Fusarium, Myxotrichium and Stysanus. These all have capacity to decompose cellulose and causing pigment staining on paper and other allied material.

Niuksha's observations in this regard indicate that Aspergillus versicolour, fulginum, pencillium, Paecilomyces Varioti, Sporotrichlum, Bombicium, S. Polyporum, T. Coningi and Trichoderma lignosium are some species of fungi having remarkable capacity to dissolve cellulose.

Flieder, who has worked on the problem of microbial deterioration of paper at the Laboratory-de-Cryptogamie of the French National Museum of Natural History in Paris states that several species of fungi grow on paper and adversely affect them. Among them are, Aspergillus, Alternaria, Batrytrichium, Chactomium, Curvalaria Epicocus, Cladosporium, Fusarium, Pencillium and Pullusaria.

Dr. S.M. Nair, who conducted studies on the isolation, culture and identification of fungi from old paper manuscripts, has mentioned that principal species of fungi found active infesting documentary wealth are, Aspergillus Niger, Asperegillus rectrictus, Cladosporium, Curvularia, Fusarium, Pencillium and Trichoderma.

Most of the fungi grow between the temperature at 0°C to 35°C but the most suitable temperature for their growth is 20°C to 30°C. The ability of fungi to withstand extremely low temperature at minus 195°C has also been demonstrated. Fungi prefer an acid medium for growth. Documents, gathering dust, lying uncared and damped are mostly attacked by fungus. Dust particles actually in rainy season absorb, moisture. Acidic gases readily soluble in water are also absorbed by papers developing acidity and nucleating fungus growth.

In the areas where humidity range is very high, fungus growth on all the organic materials including paper, wood, leather, etc. is a common feature in India. In fact, fungus spores are present in the air around us. At suitable atmospheric conditions, especially when humidity rises and lasts for a long period making the material moistened, these spores embedded over such surfaces start producing fungus cell walls. Fungus being a parasite, for its expansion always consums material from the surface on which it grows.

Mould and mildew are interchangeable terms and can be recognised by the progeny masses (colonies) of spores formed on the surface of materials. Dampness sustains the growth of mildew. Most of the moulds also cause fading of inks. It is difficult to restore the writing faded by the agency.

Vegetable, tanned leathers are more susceptible to mould growth than chrome tanned skins. Microscopic examination of mildewed leather, has revealed the fact that fungi rather than attacking collagen (a chief constituent fibre of leather) attacks the substance incorporated in leather during tanning.

The development of fungus is marked by three successive stages. In first stage, fungus attack is surfacial and can be vacuumed off or brushed easily. In second stage, spore's roots deepens the surface on which fungus grows. The third stage, is however, fata for the life of paper and archival documents. In this stage, paper infected with the agency, becomes fluffy and moistened grappled with various coloured stains. Badly infested papers are difficult to handle and give anti-fungal treatment without damaging them.

Foxing

Though, there is need to advance a very extensive research on the foxing, yet rusty brown spots, appearing on old papers is regarded as foxing. Purendu Basu, stating about the agency says that ugly brown patches are called foxing, which results from localised deposit of iron rust. Backwith observes that characteristics of foxing are partially at least associated with fungus growth but points out that it may occur due to the presence of iron in paper. Thomas Iiams and T.D. Beckwith who conducted a research on foxing, concluded that it is independent of the intensity of development of mould. Foxing stains are result of reaction taken place between iron impurities present in paper and the organic acids secreted by fungus. According to Amber and Fenny, micro-organisms are directly responsible for foxing. Their growth is sustained due to presence of glue and starch forming hygroscopic areas on paper in which water soluble products of decomposed cellulose accumulate and assume a brown red colour. Studies have brought the fact to light that fourteenth century paper which is made of almost pure cellulose is seldom foxed while late eighteenth or early nineteenth paper are extensively attacked. Rusty brown marks on the dry and printed documents having no fungus stain in general identified as foxing marks.

White Ants

The termites, potential destroyer of records are reported to exist all over the world except in Antarctica and Arctic etc. R.C. Gupta, in his article, "How to fight white ants" has reported that the term white ants the popular English name of the termites is essentially misnomer. White ants are neither white nor do they belong to the order hynoptera to which true ants belong. So far as the colour of termites, it may vary from pale yellow to dark brown. Some species of the insect, existing in Cylone now Sri Lanka, are black. Almost one thousand and half of the species are reported to exist but India claims nearly a hundred. Some of them are known as ool, ooli, sainic, karayana or valavi.

E.A. Buck has mentioned in his article "Bookworms", that termite have ruined more books than any other group of book insects. The termites are never seen running about over books and furniture unless their feeding chambers have been broken open.

Social behaviour of termites is noteworthy. Each society is consist of two parents and offspring. The two generations of the insect live in mutual co-operation in a common shelter. Their social system is so developed and organised that it seems ideal and its study is of considerable interest. Each termite colony is represented by the King, the Queen, the Sainik and the Labour. A termite colony is normally the offspring of one queen which may live upto 10 years. The soldier, the male and female are wingless and sterile and have pigmented head, vestigial eyes and a variety of mandibules. The first crop of eggs, mainly produces the workers and the nymphs are nursed by royal pair and afterwards they only take responsibility of adding population of the colony. The queen, functioning as machine, lays eggs between 30,000 to 80,000 a day. Due to sedentary life and nutritious diets the queens during the course of life acquire a grand physical structure measuring 2-4 inch length, 1 inch width and 1 inch height, about 20,000 times greater of the size of an average worker. Setting up new colonies is part of activities of kings and queens. Besides, primary queens and kings, the colony of termites is represented by secondary of substitution kings and queens namely, the brachypterous and the apterous forms respectively. In case of any mishap to the parent queen or division of the colony, the secondary queen takes upon itself the fun of reproduction of the colony.

The soldiers of a termite colony take responsibility of defence of entire colony against outwardly dangers. Procurement of food, building of nests, maintaining cleanliness, repair and tendering of eggs and nymphs are the principal duties assigned to the workers of a colony.

Depending on their habitats, the termites are divided in two main groups:

(1) Ground-dwelling termites.

(2) Wood-dwelling termites.

The former group is further classified in three sub-groups, namely: (i) subterranean termites, (ii) carton nest building termites, and (iii) mould building termites.

The major damage to building, books and other documentary wealth in Indian sub-continent is caused by sub-terranean termites. The colonies of termites are always started by royal pair borrowing into the earth. The sub-terranean forms are so called because they must maintain contact with the moisture in the soil beneath the building in which they are active or causing damage. A typical sub-terranean concentrated nest consisted of a royal domed chamber, connected by a network of galleries with numerous chambers containing fungus gardens, rests a foot or so below the ground. A large fungus chamber, adjoining the royal chamber and also a pair of main ventilation of passage running more or less vertically through the nest, commonly exists in colony.

Carton nest building termites, construct an oval shaped cellular nest of carton, a compound of earth, wood and faical matter, located either in ground or in wood or in branches of trees.

The termites capable of building mud hills of varying height are termed as mound building termites. Height of the building sometimes attains the height of even 20', having a basal of about 12' diameter. For erecting mud hills, the earth particles are joined together by the termites of this species using saliva. The humidity inside the mound is remarkably controlled and maintained at high level even in arids. Really they are air conditioned sky-scrapper of the termite world.

The group of wood dwelling includes dry wood, wet wood and powder post termites. The colonizing pair to settle colony, generally enters the trees or wood work of a building and the entire colony is confined to wood. Dry wood termites require no contact with soil or ground and may be destructive when they occur. This species is found in most part of United States but a tropical form of that is found troublesome in the U.S.A. from Charleston, S.C. Southward. They are reported to be more destructive in Cuba, South Florida, Hawai, part of California and Philippines.

Termites are voracious eaters of cellulose, abundantly found in wood and plants. Though, cellulose is difficult to digest, yet it can be done by the termites because of presence of protozoas in their intestine. Cellulose being the major constituent of paper, paste board, cloth, flanks, and other allied materials carrying information, are highly susceptible because of their food value. Termites are never seen running over furniture and books unless their feeding chambers are broken open. Sub-terranean termites' destruction in Archives and Libraries where records and books are stored in wooden shelves and containers is very usual. In private homes or in public institutions where valuable records are kept in basement rooms or even the floors of first storey, may be attacked by the agency. The documentary materials stored in wooden boxes or in the boxes made of organic material have been ruined within a time span of four months period, has come into notice. It may, however, be remembered that termites are selective feeders and much of material damaged may not be actually eaten but bored through or corroded with moisture or body secretions of the insects.

Ta!king to the habit of termites, licking or grooming the bodies of one another is a developed system of mutual feeding with saliva and body secretions. Workers relish the fatty substances exuded by the queen. Trophallaxis is another important feature in the economy of the insects under study. Even the eggs fail to hatch unless liked and tended. Ejected faeces, dead bodies are devoured. Cast skins and sickly individuals are eaten again and again till last trace of nutrition exists.

Identification of the Species

Successful control of insects in an Archives or Library is the subject of correct identification of the agency set in devastating material housed there in the repositories. The definite way of identifying a pest is to examine the adults, larvae alive or dead, their cast skins, pupae or pupal cases. The pests which leave evidence on the material they feed, usually or frequently do not remain on the damaged material or in the immediate vicinity. As such, their hide outs available in wood or other kinds of materials, sometimes require identification as secondary source for collecting evidences. Termites, woodborers, thysanura, certain dermestid beetles and cockroaches are the pests which are found on secondary sources.

The work of determining the cause of damage, when the pests have left the place, is not less than a work performed by a

Table 3.3 : Characeristic damage caused by different insects		
Material	*Characteristic damage*	*Insect Identified*
1. Surface of paper, cloth	(i) Faecal pallets of some of the species as large as those of mouse. (ii) Faecal deposits often leaving dark stains, or smears. (iii) Roughly chewed surface	Cockroach
2. Surface of paper and cloth (binding)	(i) Presence of hard irregular faecal pallets. (ii) Pieces of cast-skins. (iii) Presence of slightly itched irregular marks (iv) Presence of minute irridescent scales.	Silverfish
3. Bound Books and volumes	(i) Presence of parallel sided tunnels about 1.5 mm wide inside the binding and through the pages, running in different directions. (ii) Presence of number of circular "shot holes" on the binding and the edges of paper. (iii) Sides more seriously damaged than the middle of the bound book or volume. (iv) Gallaries are generally filled with granular dark brown dust.	Indian book-worms beetle.
4. Books, paper and wood	(i) Presence of irregular tunnels (in case of books and manuscripts well stuck with mastic like pass). (ii) Burrows on wood, books and paper lined with extrement and earth cemented together.	Termites (Sub-teranean).

detective. Some damages can be identified at a glance while more often it may necessitate to hand-lense or microscopic examination.

Obviously, the damage recognition is a very important part of pest control. The exit holes made by insect may be of characteristic importance. But in case of pests when they found in huge numbers, the evidence like faecal pellets, scales faccal stains, webbing, mud tunnels, hairs, etc. prove to be of significance in the job of identification of insects. A careful note of all these factors, often makes possible to identify the insect correctly. Following characteristic damages, have been used by the custodians and conservators for identification of insects before planning strategies against insect devastation.

REFERENCES

Agrawal, O.P., "Environment & Preservation of Paper", *Conservation of Cultural Property in India,* Vol. X, 1977, p. 15.

Buck, E.A., "Bookworms," *The Indian Archives*, Vol. I, No. 2, 1947, pp. 126-34.

Basu, Purendu, "Enemies of Records", *The Indian Archives,* Vol. IV, No. 4, 1950, pp. 7-13.

Chand, Punam, "Control of Termite", *Conservation of Cultural Property in India,* Vol. IX, 1976, p. 12.

Gupta, R.C., "How to Fight White Ants", *Indian Archives,* July-Dec., Vol. 8, No. 2, 1954, p. 122.

Hughes Mckenny, A.W., "Protection of Books and Records from Insects," *The Indian Archives,* Vol. 2, No. 1, 1953, pp. 41-43.

Hot, Chand, "Ecosystem and Fungal Growth, Conservation of Cultural Property," *Newsletter,* No. 15, Jan., 1988, p. 9.

Kimberly Arthur, E., "Recent Development in Records Preservation," *The Indian Archives,* Vol. III, No. 1-4, 1949, p. 70.

Mathur, D.K., "Protect Your Archival Material," No. CH/NCE to Cockroaches—Conservation of Cultural Property, *Newsletter,* No. 15, Jan., 1988, p. 11.

Mitra, Rajendra Lala, "A Report on the Indian Manuscripts in Indian Repositories," *The Indian Archives,* Vol. 2, 1948, p. 33.

Nair, S.M., "Biodegradation of Paper, Conservation of Cultural Property in India," Vol. X, 1977, p. 13.

Nair, S.M., "Some Simple Methods for Examination of Organic Museum Objects for Insect Attack, Conservation of Cultural Property in India," Vol. XI, 1978, pp. 2-29.

Prajapati, C.L., "Archivo-Library Mateials, Their Enemies & Need of First Phase Conservation," at pp. 77-97.

Reichardt Gunther, "The Durability of Paper", *The Indian Archives*, Vol. I, No. 3, 1947, pp. 230-32.

Roonwall, M.L. and Chatterjee, P.N., "The Control of the Indian Book Worm beetle Gastrallus Indicus Reitter, Indian Forest Records, Entomology," Vol. 8, No. 6, 1952, pp. 53-54.

Roy Guha K.K., *The Indian Archives*, Vol. I, No. 2, 1947, p. 124.

Shaheen Fauzia, "The Insect Factor in Deterioration of Wooden Cultural Property, Conservation of Cultural Property" *Newsletter*, No. 15, Jan., 1988, p. 12.

Singh, S.P., "Particulate Matter in Air and Works of Art, Conservation of Cultural Property in India," Vol. X, 1977, pp. 57-60.

Thomas M. Iiams, "Preservation on Books and Manuscripts," *The Indian Archives*, Vol. I, No. 1, 1947, p. 39.

4

MEANS AND WAYS ADOPTED FOR PRESERVATION

Conservation of documentary wealth is an universal necessity now. The question is why? The simplest answer to the question is that they are prime source of information which has played a pivotal role in world building and development of human society through the ages since evolution of mankind. In other words, the records constitute the tangible memory of organization, society, nation and universe as whole reflecting the vivid activities of the period of their creation and hence need all protection ensuring their physical defence. I cannot do better than quoting Purendu Basu, an imminent Archivist of National Archives of India and other who emphasized over the necessity of preservation (Preventive Care and Curative Care). Basu says that preservation of records is necessary as they have value attached to them. In words of Philip C. Brooks, the records are the tools of administration, the embodiment of experience, protectors of legal rights and source of many kind of information and hence need preservation. Summing up the whole gamut, a document of Archive character, whatever other qualities it may possess, is first and foremost, a vehicle of impartial evidence. The records indeed may possess other qualities; it may be microscopically small or monumentally large, it may be thousand or few years old, it may be object of matchless beauty or an offence to the eye, its written contents may be comic or dull, trivial or revolutionary but all these qualities are secondary to the basic fact that a document conveys, a piece of evidence and source of information and on this account it must be conserved against decay, damage and spoilage.

It becomes almost clear from the studies in Chapter-3, that documents being organic in nature are bound to decay with the passage of time. This process of decay is governed by the elements of physical and chemical nature. Further, the documents containing the materials of food value for some insects are damaged by them. Careless handling and neglect by the people, generally also results in their damage. Besides, some agencies bring about spoilage to this wealth.

In view of the above, to keep the documents safe and sound for the use of posterity, their preventive conservation is a must. The ages old proverb, prevention is better than cure, still holds good in the field of documentary conservation. Preventive and curative care are the two elements of preservation. If preventive measures are effective and well executed, the necessity of curative treatment can be avoided for a long.

The preventive care helps in preventing decay, growth of biological agencies and also preventing wealth from eventualities of occurrences of any man made and natural calamities. The essential elements for preventive conservation are:

1. Suitable building facilitating ideal storage conditions and scientific storage.
2. Responsive fire prevention, detection and controlling measures.
3. Preventive measure against biological agencies.
4. Safety and security measures.

Suitable Archival Buildings

Suitable archival building is the basic requirement of modern conservation. Without it, neither the records can be preserved against various enemies nor safety and security measure are ensured. Purendu Basu, in his article. 'Enemies of Records' has stated that from view point of preservation and security, a specially constructed and air conditioned building, is essential.

The requirement of a modern Archival functional building

however, could be divided into two: (i) the requirements from the view point of preservation, and (ii) the requirements facilitating disposal of works efficiently and conveniently.

The first part of a building, involves suitable site, design, orientation, load bearing capacity, provision for prevention, detection and control of fire, effective light control system, ideal storage conditions, dust removing provisions, proper air circulation, up-to-date storage facilities and good housekeeping.

The second part of the building needs planning that the various sections, involved in performing various activities, such as reception, fumigation, repair, photo-duplication, research are so placed that they ensure disposal efficiently and conveniently. For the first time in 1964, to fulfil the requirement of Archives, Bureau of Indian Standard (Indian Standard Institute then) prepared and published a code of Practice Relating to Primary Elements in the design of Building for Archives. The typical flow diagram for movement of Archives published in the standard can be seen in the Fig. 4.1.

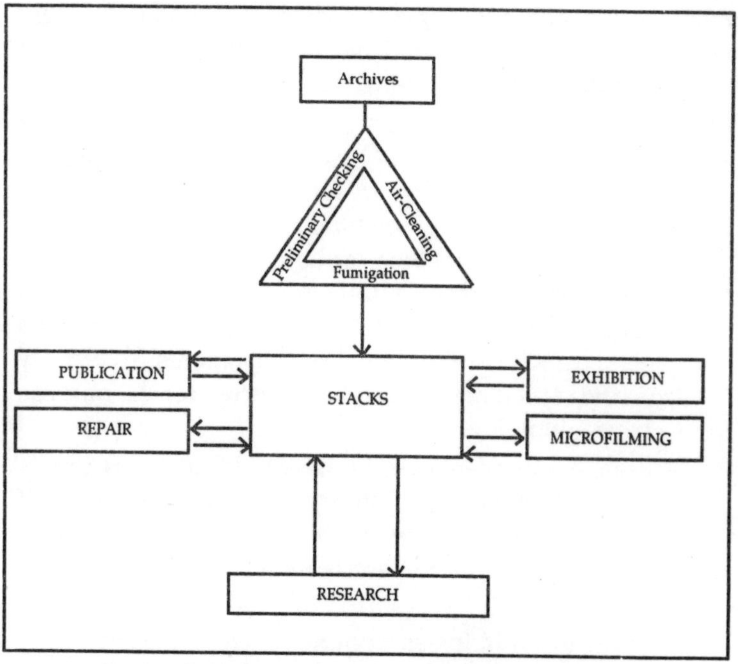

Fig. 4.1 : Typical Flow Diagram for movement of Archives.

The negligence of any provisions required for suitable archival building will not only hamper the preservation of records but also restrict the activities to be carried out for the purpose as well. Remember, building is a basic tool for conservation of documentary wealth. As such while considering the case of building, the suitable requirements for both the categories of building i.e purpose built and adopted are must to be accounted for. Some changes in latter type of building, suiting to the requirement of preservation are must to be incorporated while in previous category, all the requirements with future plan, should be made available to the architect preparing plan and designing.

The first and foremost requirement of an Archival Building is suitable site or location.

Site

Suitable site of an archival building in one hand minimizes the risk from ravages and on the other decreases the load of expenses to be incurred for conserving the wealth. Besides, architecture of the building and materials in its making also need consideration from the stand point of preservation. For selecting suitable site, following points are to be kept in the mind.

(i) Land-sliding is a big problem in hilly areas. As such if exigency of establishing an archives or Library in such areas arises, the site chosen for the purpose be in the safest zone having no possibility of land-sliding. It is ensured by the geological records or the happenings of the past. If no such zone is available in the region, ignoring the general principles of Archival establishments, a site away from the capital place should be selected for the purpose to have a building.

(ii) Termite is a problem in some areas. Such areas are totally unfit for Archives and Library buildings. However, if necessity persists to have building in such areas modern facilities to have a termite proof building should be well availed.

(iii) In order to avoid losses from flood, etc. the bank of any river, *nalah,* canal and low land which had ever been the inundated area, should be avoided to have the buildings in question.

(iv) Land made of sub-soil, damp land levels to be effective of heavy seas are not considered fit to have an Archival building.

(v) The areas, having history of frequent earthquakes are unsuitable for Archives and Libraries as they also bring about considerable loss to the documentary wealth.

(vi) Sites near Chemical plant-stations, store-houses causing intrinsic dangers such as fire, a place near petrol tanks, explosive depots. gasometers, air fields, railway stations and major bus stands are not fit for Archives and Libraries.

(vii) Pollution brings about serious damage to the documentary heritage is a well known fact now. As such, to save the wealth from noxious gases, the buildings in the proximity to congested residential and industrial localities is not suitable at all. The site chosen should not only be free from air pollution but also provide calm and pleasant surrounding conducive for academic work.

Design and Orientation of Building

The design and orientation of building is planned keeping in view the local problems and to maintain a natural intake of air. Building materials providing insulation against external heat is advantageous. The designs of some of the monuments of Rajput and Mughals Architecture may be a suitable design for Archival building. In modern context, of atomic wars, such architecture once again has proved its significance for archival security.

The insulating and flooring materials getting place in modern buildings, have been the source of indoor pollution. As such, while going to make use of such materials their survey from stand point of preservation is necessary.

Energy consumption in the form of electricity for air conditioning is the serious problem of Archives in general. To sort out this problem, Natural Air Conditioning System, has been introduced in Bundes archives Germany. Possibility of such buildings in Indian subcontinent and African regions needs planning and study. Once it is found suitable revenue of millions could be saved annually reducing the financial burden in Archives.

General Requirement of Archival Building

1. The entire rung of the building facing North and South in stead of East and West, is considered good orientation as it avoids direct entry of sun-light in the building especially in strong-rooms.
2. If possible, the surrounding area of building should be made vegetated to suppress the effect of pollutants, dust and heat.
3. Sufficient space should be available all around the building to carry out fire fighting activities without any obstruction at the time of any serious fire accident.
4. A building should have a responsive ventilated system and capable to obstruct the entry of outdoor sun radiations, hazardous to records.
5. The floor height of a building should allow racks of the height facilitating search cases without use of stool, ladders, etc. leaving a space of 25 to 31 cm between the top frame of racks and ceiling for free passage of air. The floor height in European countries, in general, varies from 2.10 to 2.15 m, but in tropical and sub-tropical countries, this height should be more depending on the rack height.
6. A building should be well equipped with modern facilities of fire prevention, detection, and control system.
7. The weight generated by records normally varies from 50 kg. to 60 kg. per linear meter. But in special cases this variation extends from 60 kg. to 70 kg. per linear meter if records on thick paper are created and stored. This weight per linear meter increases more when compactus system of shelving is introduced. As such, the plan of building and its structural design should be so prepared to meet the above requirements, failing which building structure may collapse.
8. If building is not isolated, focus is to be attended to the possibility of flames reaching the buildings at the adjoining, incase of any fire incident.
9. Care should be taken to select materials for construction

of building that are of no fire hazards and do not promote deterioration of records. Wooden partitions, furniture, windows, tapestry, inflammable paints and varnishes all add to these.

10. Strong-rooms should have sufficient windows and ventilators fitted with water shades and sub-brokers. It is essential to maintain air circulation and restricting entry of rain showers. The size of strong-room neither be much larger nor smaller. The suitable size ranges from 150 sq. m. to 200 sq.m.
11. A building should be ready to adopt central air conditioning system without major changes.

The another part of functional archival building deals about the placement of various sections facilitating disposal of routine duties. In 1947, on this aspect light was thrown by S. Chakravorti, in his article, "Virginia state Library" and the subject matter was made perspective through relevant plates relating to the library.

The system adopted in National Archives of India, seems to be very identical to the system adopted by Virginia state library, which was a model for small states in USA and abroad then. Some common features of both the buildings establish the above fact. Some elements of similarity in Virginia state library and National Archives of India are, the nomenclature of stack areas and placement of sections. The ground floor of the state library in USA is called first floor and in N.A.I. also ground floor is so called. Fumigation and cleaning room (preservation room), the photographic laboratory, the rare book room and part of stack, (see plate 4.1) get place on the basement in the Virginia state library. In N.A.I. these sections, with slight additons are located on the ground floor. First floor plan and plan of stack area of Virginia State Library are available in plates 4.1 and 4.2 respectively.

Planning Strategies Against Fire

Focusing attention on degree of loss, brought out by fire, it can be stated that fire is such an enemy which does not leave any thing but charcoal and ash.

Paper is fire promoting material. Once fire outbreaks, it

spreads rapidly. In order to meet out challenges from fire, a modern Archival building should be equipped with fire prevention, detection and control systems. R.C. Gupta, in his article, "Building and storage equipment for long preservation of paper", has recommended that the building intended to house the permanent collections should be three hours fire resistant and provided with fire resistant materials.

To save the wealth against fire, measures for prevention, detection and fighting fire, with provisions of evacuation in building, are direly needed. While planning strategies for preventing fire the possible sources causing it may be identified. Smoking, and short circuiting are the well known possible sources. As such, the first step to be taken is to check smoking in the repository and nearby areas. The independent sectional wiring system through conduits and isolation of one strong-room from another, is time tested system to control fire. Besides, use of standard electrical wires and fixing of all electric fixtures on the walls or roof i.e. pendant type electric lights should strictly be avoided. For use of electric appliances like heaters, hot-air blowers, vacuum cleaners, air circulator, etc., sufficient number of electric out-lets should be provided to obviate use of long flexible wires. All the repositories should have main switches outside the provision of pilot lights and these should be switched off after office hours.

Inspite of effective preventive measures, it cannot be ensured that no outbreak of fire would occur. The quick and effective detection system in an Archival and Library building is of great help in controlling the fire and preventing losses. The old tradition of relying on human vigil round the clock for the purpose of detecting fire is not dependable. Automatic detection systems, making use of all the ingredients of fire, (light, heat, and smoke) raising alarm are available now and they should be made part of functional archival buildings. Heat, smoke and optical detectors give the right answer to the problem of detecting fire. "The view system monitored by lasers and the system based on change of air quality are more advance systems for detecting fire."

After the fire is detected the next step is the fire fighting. For this purpose, various fire fighting aids such as paints, blankets, and other alternative are available today. But it requires judicious selection, keeping in view the safety and security of records and other documentary wealth.

Soda-acid type, carbon-dioxide type, foam type, carbon tetra-chloride type, powder type and water-type, are some old varieties of fire fighting equipment. Now system of water reservoir making use of wet and dry rider system and halone type have also come up for frequent use in suppressing fire.

Every fire fighting system works on the principal either to bring down the temperature of substance under fire, below ignition temperature or to cut off the supply of oxygen from the surrounding or to remove the substance from fire or all of them. Oxygen supports ignition. If its supply is barricaded, the fire will automatically cease off.

The main product of soda-acid type[+1], and foam type[+2], fire extinguishers is carbon-dioxide. But sometime, acid and water is also carried with the powerful blow of carbon dioxide which is produced by the extinguishers after it is made operative.

All the extinguishers making use of carbon-dioxide or carbon-tetra-chloride, when operated in the fire affected places, a huge amount of carbon dioxide spread over the area and suppresses the percentage of oxygen. Due to inadequate amount of oxygen, intensity of fire gradually comes down and finally dies off.

Powder type fire extinguishers make use of powders forming foam on heating. When such powders are poured over fire hit objects, a foam is formed which covers the objects and cuts off their contact from the surrounding environment. Because of it, the supply of oxygen from the ambient to fire hit objects is cut off. Thus, fire is controlled. But foam so formed sticks to the fire hit substances which cannot be separated easily. Water on the other hand, brings down the ignition temperature of the substances and kills fires.

The use of powder type and water based fire fighting systems, no doubt, has increased today in controlling and combating fire. But these are not suitable at all to the Archives and Libraries because of the reasons that separation of

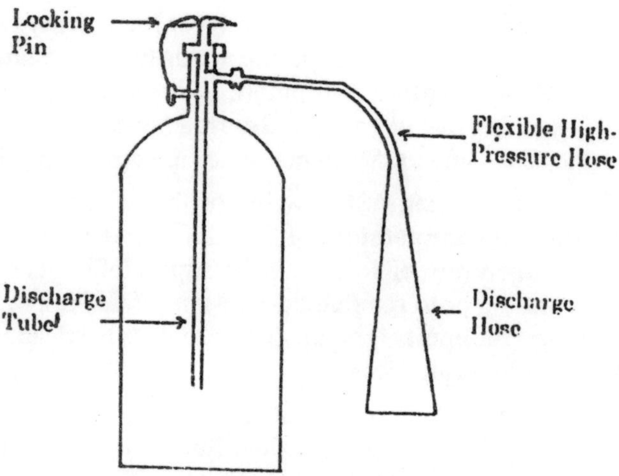

Fig. 4.2 : Carbon Dioxides Type Fire Extinguisher

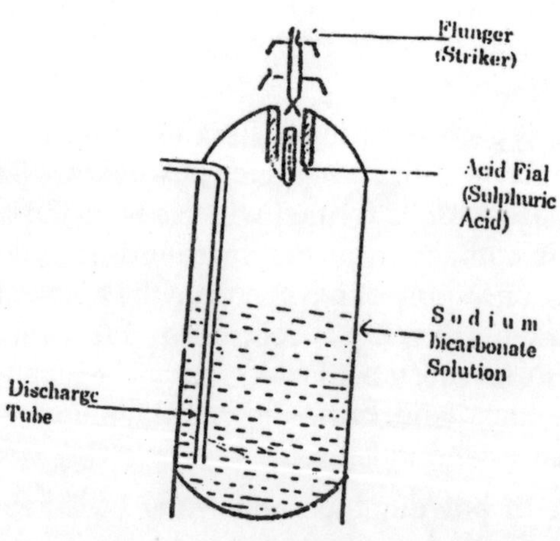

Fig. 4.3 : Soda-acid Type Fire Extinguisher

documents from the foam without any damage is not possible. Secondly, water combats the fire but at the same time acts an wetting agent. Once paper absorbs water, its mechanical strength falls down and the water soluble ink bleeds from the document. No doubt, in case no alternative to control the fierce fire, is available, it can be used but normally should be avoided. Carbon dioxide type and foam type fire extinguishers, anti-fire blankets are the best choice for Archives and Libraries. Fire resistant varnishes in doors, windows and stacks help in controlling fire. But such varnishes before use should be tested for their suitability on documents. If they cause for indoor pollution, their use should be discouraged.

Besides, a hot line service between Archival institutions and local fire equipment dealers, periodical checking of detectors and fire fighting equipment, readiness of fire team and trained staff are the dire needs of Archives. To keep the system alive and people alert towards any eventuality of fire, fire-drills should be organized from time to time.

Effective Light Controlling System

It is well known fact that ultra-violet radiations (below 3600 Å) cause direct rupture of cellulose bonds. The ultra-violet radiations generated by natural source of light (Sun) and fluorescent light may have their presence in the stack areas. As such, the effective light controlling system involves two basic things, i.e. the restriction of entry of invisible radiations from outside the building and the filtration of light emitted by sources of light installed in the storage areas. To prevent entry all the possible sources (windows, and ventilators) where through light radiations get entry are covered with tinted glasses or glasses containing oxides of titanium, cobalt, or iron. The colours effective to absorb ultraviolet radiations are red, green, yellow, blue and black reports V.V. Talwar in his studies. Moreover, draping of cloth curtains of either colours referred to above in windows is considered to be of additional advantage. Use of special ultraviolet filtering materials such as UF-3 plexiglas or acryclite op-2 in windows or fixtures, facilitates to some extent in filtering out U.V. light.

To achieve the required illumination (75 lux for paper 75 uw/lumen for micro forms) in the stack area, filtration of light emitting ultraviolet radiations is a must and that can be obtained by slipping ultraviolet plastic/glass sleeves over fluorescent tubes. Moreover, if required, suitable lux of (7.5 lux) light in the area is obtained by installing less number of tube lights. This can be achieved by smoothering walls using quality light reflective paints on them. White paint (titanium dioxide pigment) on the walls and ceilings of the rooms is effective to absorb near ultraviolet raditions. Near ultraviolet radiation from both, natural as well as artificial sources, are absorbed by the paint upto 90%.

Ideal Storage Conditions

Dry, ventilated, dust-free, and adequately lighted accommodation is considered the suitable place for preservation of documentary heritage. Besides, optimum temperature (20° to 25°C) and humidity (50% to 60%) and unfluctuating environment are also make part of constituting ideal storage conditions promoting preservation. To maintain the latter three, the right answer is central air conditioning. In terms of conservation, no system will give the greater return than a controlled environment at optimum ranges and well supervised programme of house keeping.

The four functions of air conditioning are ventilation, filtration, humidity and temperature control. By filtration, it is possible, to clean urban air for most of its contaminating noxious gases and acropolis, minimizing the grime abrasion and chemical deterioration. "For optimum efficiency in an air-conditioned building, atleast 5-6 air changes per hour with atleast 20-25% fresh air intake per cycle is necessary to minimize the contaminants added by people and material to the air."

For conservation of Archives, a humidity range above 60% encourages mould growth, and below 45% makes paper and papyrus brittle, warps wood, turns vellum horny and leather to behave unpredictably. However, if air conditioning system is introduced, the due care for preventing fire in the building, must be taken care of as air conditioning ducts, lift wells, chimneys and dumb waiters not only provide passage for fire spreading from one floor to another but also act as a drought. Better is the system to have independent air-conditioning ducts for each floor. In case, if they run from one floor to another, they should pass through fire resistant walls having provision of fire and smoke dampers. The return ducts should be fitted with smoke detectors suitably coupled to ensure complete shut down of conditioning plant at the event of outbreak of fire. The plant room should be isolated from stack areas etc. by 3 hours fire resistant walls and should have access from outside.

The central air-conditioning is the most effective and Practical way of controlling temperature, humidity and cleaning air but a costly affair, normally out of the reach of small units of Archives and libraries. In such cases, manual system of controlling temperature and humidity at optimum ranges should be used.

The measures to bring down temperature, in non-conditioned buildings includes the placement of water at convenient places of stack area in shallow pots, use of air-circulators and exhausts etc. The temperature and humidity behaves in inverse proportionality. Means when one increase another decreases. Specially, in summer season when relative humidity in the area decreases, temperature rises. Conversely, increasing the humidity in the areas (not above 60%) the intensity of temperature could reasonably be lowered. Humidifiers are the right answer for this purpose.

High humidity, can be reduced by employing chemical and mechanical methods. Silicagel is the time tested chemical. Normally, 2.5 kg. of silicagel is sufficient to reduce humidity in a closed space of $25M^3$. Quality silicagel (SiO_2H_{20}) when attains saturation absorbing water, its colour changes from blue to light pink. Saturated gel may be reactivated by indirect heating to 120-160°C. However, during reactivation water contents of gel should not go below 10%. Monitoring of the work is the essential tool to preserve the wealth effectively. Silicagel is placed near windows, doors and other places in suitable containers bearing holes or shallow pots.

Dust Free Stack Area

The dust in tropical and sub-tropical areas poses a serious problem to records bringing about their physical as well as biological embritlement. It is common practice that files after closing are transferred to a record room, where they get a permanent place in open racks or open wall shelves till they are handed over to Archival depository. Hardly care is made to tie them in bundles and to drape curtains on the open shelves accommodating files. As a result, dust particles deposit forming layer over layers on the files, rain showers make them wet entering through ventilators and windows, direct sun light falls over them, but no one bothers about the precarious condition of records. Let the record enemies grow, multiply and launch their life activities freely but no one is going to disturb and prevent them not to make their permanent shelters in records.

Obviously, for preservation of records, It is must to save them from hazards of dust. For the purpose stack area should be kept dust free. It requires three fold planning i.e., restricting entry of dust from outside, preventing dust creation inside the building and dust extermination.

Ensuring prevention of dust entry, windows and ventilators, in the area, after office hours and during working hours when fast wind and storm blow, should be shut off. Curtains should also be draped, during these hours. Old plaster material causing dust should be replaced by new one. Filtering devices to clean urban air of most of its contaminated aerosols are considered very effective. The accumulated dust over the records, shelves and other materials should periodically be removed using vacuum cleaner which is the most effective cleaning device today. In case of numerous stack areas, a suitable plan for removing dust in a short period of time should be chalked out. Wiping system using wet cloth, jute may also be employed for cleaning floors and shelves from time to time using traditional system instead.

The compressed air guns and ionised air guns are considered effective aid for custodians of Archives and Librarians in their never ending problem of dust.

Proper Air Circulation

Stagnant air in stack areas give rise for growth of insects and fungus. From the stand point of preservation, therefore, it should be ensured that no stagnant air pocket is formed within the area. For the purpose proper air circulation in the area is ensured which is augmented with the help of air circulators, fans and exhausts.

Storage and Storage Materials

Preservation of documentary heritage needs not only the scientific storage but also the storage materials of the quality. There was lack of scientific methods of Archives preservation in olden days. People used their own methods of storage based on experiences to save them against ravages, for longer period.

There was a system of keeping manuscripts wrapped either in leaves or barks or cloths. Tradition was also there in Indian region to put them in hollow cylinders made of bamboo or any other wood as per convenience. Subsequently, a system of hanging rolls of manuscripts under roof, generally used for cooking or for ritual of Havana also started as smoke was capable to prevent them from biological attacks. Today these methods, in modern archives are not in vogue, firstly, because they are not practicable. Secondly, diposited smoke particles creates other problems.

Manuscripts like *Patiya* and Palm leaf were gathered one over the other and were tied with a soft rope passing through central hole over them. Such bundles were stored or hanged in the rooms specified above. With the advancement of time a system of tying manuscripts between two hard wooden plates of moderate thickness was also introduced. Three prominent formats of storage i.e. Bastas, bundles and containers are very common today. They are effective for scientific storage.

The manner in which documents are shelved and used had a deleterious impact on the preservation. Untidy shelves, rough handling of records lead to damage them and in a needless waste of resources. John Davis observation, mentioned below, is an evidence of the fact stated above.

> "It was common practice in most Government agencies in the past, to relegate the older or non-current records to basements or attics of the office buildings when space was required for current records. The characteristic feature was when an office could no longer house its old records in the registry, it was an accepted practice to remove them out into a nearly store-room in which the office stationery was kept. At times, when the first store-room was full, the dead or non-current records were cast into some remote store-room where amongst old tyros and scrap

iron, they would lie forgotten and neglected in filthy and decaying heaps on the ground. Records salvaged from the store are now found in archival institutions where large sums of money and energy are being spent to rehabilitate them so as to make them available for research and reference."

About careless storage of records in creating agencies, attention is invited to the fact that in general, files are gathered one over the other randomly on the tables of dignitaries waiting their turn for disposal. Sometimes, it is difficult to make distinction between gathering of bundles of straws in a place and files on the table. At some occasions, files are thrown either down the table or away from it by the official just imparading the gravity of officialdom after putting note or singing them.

Preservation of Documents and Manuscripts

Throwing light on the mode of preservation of manuscripts, Shri Rajendra Lala, has mentioned in his article, "A Report on the Manuscripts in Indian Repositories" published in the Indian Archives, Vol. II, 1948, that different modes for preserving manuscripts were being adopted by different people and religious sections. A dry masonry room in general having sufficient number of shelves or chests was provided in the houses of rich people for storage of manuscripts in bundles. Hardly the wealth drew attention of people to open the bundles, ever now and then, and to expose them to the sun for few hours to make them dry. So far as storage of manuscripts in Monasteries, the same mode was adopted but the bundles were opened, aired and dried more frequently by monks taking delight in reading. The Jains were seemed to be very particular in preserving manuscript wealth in their Monstaries adopting the modes mentioned above. The case of preserving Tols of Bengal was, however, very difficult.

The men owning the wealth were not from the rich strata of the society but were very poor. They had to live in low damp thatched huts of the meanest description. They had no means to purchase proper cabinets for storage of their professional duties and frequent peregrinations to distant places for earning means of their livelihood, that they hardly can pay attention to look after their bookish and manuscripts wealth. The storage place, they usually assigned for manuscripts was a bamboo frame placed across the beams of their huts. It was exposed constantly to the damp emanating from the daily washed mud floors of their rooms and occasionally to leakage from ill made and old thatched roofs. The mice and other vermins have free access to manuscripts day in and day out.

Storage in *Bastas*

This system of storage was probably adopted in medieval period to keep records of courts, systematically. It is still being used in various record rooms, of courts and even in state Archives of this country as an important tool for storage of records, books, manuscripts, and files. For storage in this form they are wrapped in a piece of cotton cloth. Mostly, cloth of two colours i.e red and yellow was preferred for wrapping the records and books. These two colours have nothing to take with religion but to emphasize the use of cloth of two colours for protection of religious *Grinthas,* it was associated with the religion. The fact is that the two colours not only discourage the insects but also absorb ultra-violet light is a scientifically established fact now. The additional advantage of wrapping record materials in cloth is that it protects them from ravages of dust. This method of storage, however, is not suitable for long term preservation as poor documents at the bottom and all the documents from corners are damaged because of handling of *bastas.* To avoid this, a modified system in which a hard sheet larger than the size of documents should be kept on the bottom and then documents/files and entire assembly be tied in the form of a *Basta.* The cloth of either red or yellow colour should be preferred.

Storage in Bundles

Storage of archives in the format of bundle has emerged as a very important tool for preservation of records in modern Archives and record offices. All the records/files are tied between two plywood boards larger than the size of documents and shelved flat. In general, a 5 ply-board, having two grooves along each length side as shown in the Figure 4.4 is used for tying bundles. Shelving of bundles, vertically is not safer to keep up their physical defense. They should be shelved flat as shown Fig. 4.4 (a).

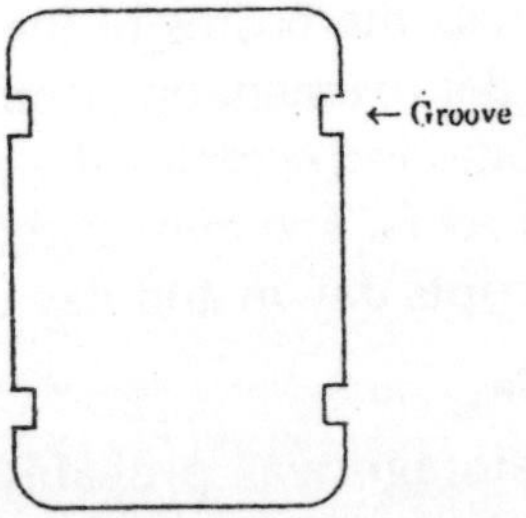

Fig. 4.4 : Plywood Board for Tying Bundles

Fig. 4.4 (a) : Storage in Bundles

Storage in Containers

Loose records may conveniently be stored in containers. A carton box made of acid free single ply board paper and cloth stronger enough to carry the load at the event of storage and shiftment is a suitable container. It is being used in Indian Archives, from many years. Since stagnant moistened air in the boxes many adversely affect the documents stored their in, it is essential to prevent it. It can be achieved by making small holes in the carton see Fig. 4.5. The characteristics that ideal carton box should have, are:

(a) Proper design, facilitating easy handling in closed and opened positions, maximum utilization of space and smooth interior and exterior of container.

(b) Fire resistant character to protect the contents against fire.

(c) Light in weight, durable and permanent.

(d) Impervious to dirt and dust.

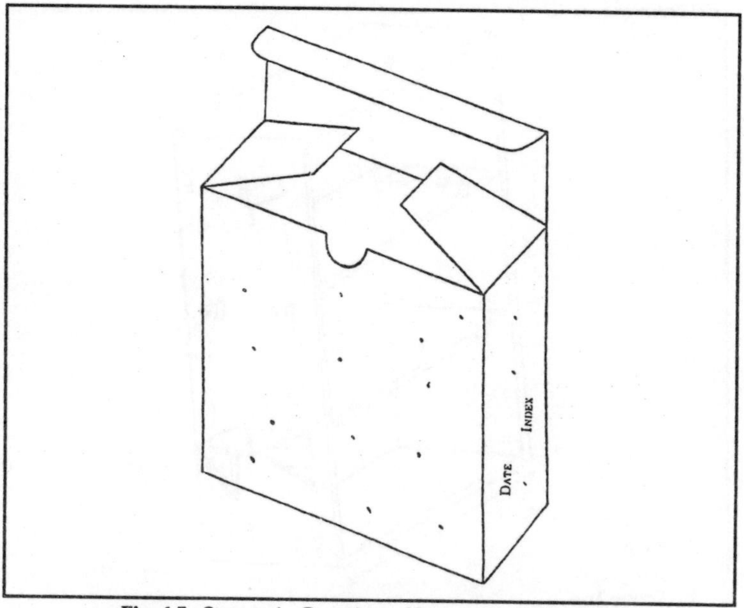

Fig. 4.5 : Storage in Containers (Carton Box with Holes)

(e) It should allow the circulation of air inside the container to prevent formation of stagnant gases.

Shelves

Material in making of shelves and shelving system are the most important aspect of documentary preservation. Maximum utilisation of storage space and convenience of access apart from preservation are the most desired objectives behind shelving. These are achieved by right selection of design and size of stack including material in its making and the system in which they are arranged in the strong room.

So far as the modern stack, its design and parts, it is comprised of 10 elements as show in the figure 4.6 to fulfil the requirement of modern Archives and Libraries.

General Characteristics of a Stack and Shelving

Since the bundles, *bastas* and carton boxes are the units of storage which are accommodated on shelves, the depth of a shelf has considerable value. It normally varies from 30-50 cm. according to dimension of documents varying from 20 x 31 cm. to 29 x 46 cm.[++]

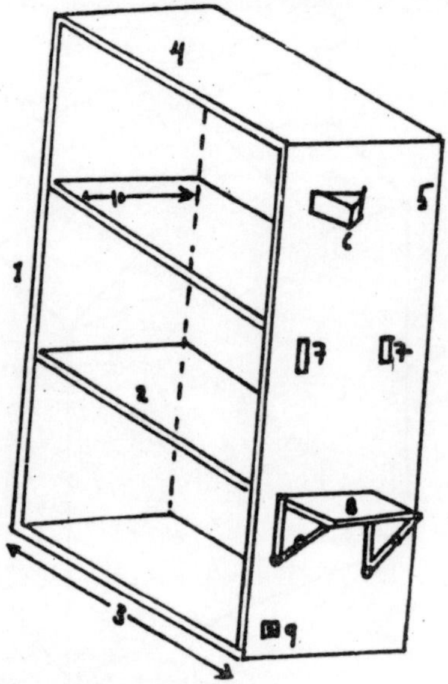

1. Upright
2. Shelve
3. Bay
4. Frame Top
5. Solid end panel bordering passage way
6. Stack identification table
7. Ticket Packets
8. Collapsible Consultation shelf
9. Electric power point
10. Shelf depth (0.2 m to 0.3 m)

Fig. 4.6 : Double Entry Shelving Stack

As such, a shelve with 30-46 cm. Depth is suitable for archives storage while for libraries, this requirements is 20 cm. The general characteristics of a stack are safety, strength and convenience. The shelving system should be easy in access, movement of records, retrieval and promotive to preventive norms.

Safety

Safety and security are the prime needs of documentary preservation. As such, a stack and shelving should be so designed and arranged that in no circumstances, it adversely affects the records. Though, rust free metal stacks are today preferred world over for storage, it has a disadvantage of condensation of moisture between the contact margin of records and shelf in the non-conditioned Archives and Libraries located in the regions marked with high humidity and temperature.

To minimize the risk, racks with perforated shelves should be preferred rather than plain shelves and focus should be attended to keep the area well ventilated.

Strength

Strength of shelves and uprights are 20 chosen that may bear load of records without showing any sign of deformation. Further it should not increase load on the floor. Documents normally generate a load of 50-60 kg. per linerar meter of Archives but this figure rises 60-70 kg. If documents have been created on thick, heavily loaded paper.

Convenience

The size and shape of stacks are chosen to fulfil the requirement of today and accommodating the holdings in the repository. Today's, racks differ from traditional racks. This change has occurred because of requirement. The various elements of modern rack has already been explained earlier. However, a bent groove in the front side top frame is useful for storage of rolled maps and out sized documents. Selection of material for making of racks has been prominent matter of convenience.

Materials for Racks

Wood has been widely used for making of racks because of convenience as it was easily and abundantly available all over the world on cheaper rates and could be framed into desired shape without much labour. But because of some demerits from the stand point of preservation of records and also from the view point of permanence, it was replaced by iron.

The glaring demerits of wood shelves are:

(i) *Hygroscopic* character, which causes dwelling and distortion in humid climate.

(ii) Combustibility

(iii) *Sagging character*: Because of hygroscopicity wood shelves absorb water from the ambient improving flexibility a little bit. Due to load of the records placed over shelves they bend downward causing deformation of rack frame.

(iv) *Prone to Attack by Termites and Insects*: Since wood contains the material of food value for termites and wood borer, it invites them. If termites attack, their activity will not be confined to wood but also extend to records. Wood borer make holes on the racks which are subsequently used by other insects found in records as their shelter places causing trouble in their annihilation.

The right answer to avoid all demerits mentioned above is steel. This is the reason that steel racks are preferred world over today. However, one disadvantage of iron is that it gets rusted which migrates to the paper material kept in close contact. With the result documents are acidified. It can however, be overcome by coating metal iron by suitable rust free, non-sticky paint.

Shelving

Display and placement of shelves in a systematic order in a strong room is known as shelving. It should ensure proper air circulation in the area, retrieval without irritation, carrying of records and cleaning without any inconvenience. In every prevalent system of shelving whether it is cantilever (traditional) or compactus, it should be ensured that natural cross ventilation of the strong room is not barricaded because of shelves. To facilitate examination of walls and also to carry out dust cleaning drives, a gap of 30-31 cm. between the rack and wall should be left. A gap of 30 cm. between the top frame of rack and ceiling is helpful for easy passage of air specially in tropical regions. A gap of 0.75 M approximately also be left between two successive rows of racks, to carry records and attend search cases. This gap is technically known as gangway. Big strong rooms more than 150 sq. meter however, require provision of main gangway (1.40 to 1.50 M width).

In cantilever system of shelving (traditional), the arrangement of tracks is so made that entire load of material including racks rest on the iron pillars founded on the ground. So far as shelving, racks are displayed on double entry basis. The display has been shown in Plate 4.3 and 4.5.

The shelving system at 'b' responds to all the requirements of shelving for preservation of records while at 'a' the cross ventilation is obstructed by the rack rows which is not desirable at all. The sides of a room provisioned with windows are to be left un-obstructed responding free air circulation.

Compactus Shelving

The two systems of compactus shelving, the forward movement and the side movement are being used world over today. The metal racks fitted with wheels on the bottom, moves on the metal tracks displayed on the floor of the strong room in both the systems. Conveniently, racks are hanged from the ceiling of the strong room using pulleys and wires, if they are not fitted with wheels on the bottom.

The idea of forward movement shelving system can better be understood recollecting to memory, the system of unshunted wagons on the railways, presuming Wagons as racks fitted with wheels.

Compactus shelving ensures protection against dust and light. It is economic in space saving converting stack areas into compactness. The density of storage can be increased without the expansion of building. In this system, it is possible to convert stack area into a visible security area as it is possible to incorporate a locking system. A strong room of dimension 14 x 10 m if equipped using traditional shelving of 60cm depth (double entry) having gangways of 0.70 m width and maintaining a gap of 31 cm. between shelves and wall, only 180 shelves can be accommodated while in forward and sideways movement systems, 360 and 270 racks respectively can be accommodated. Obviously, the space saving in both the latter systems is comparatively more than traditional system but the maximum saving of 100% can be achieved only in forward movement system.

Compactus system, however, is not free from demerits. The system is obstructive to free air circulation, promoting fungus growth. Secondly, the search cases simultaneously in two stacks at the same time cannot be attended. It affects quick disposal and may be frustrating. The system is not free from possible risks of distortion and jamming. The system is not advantageous to the units possessing dismal collection and lacking trained professionals of the field. Film Archives Pune and Departmental Record Room of Ministry of Planning are making use of forward movement compactus system in India. See Plate 4.6.

Storage of Books, Vol. and Outsized Documents

Registers of separate items on paper with or without seals, maps, charts, plans (out-sized) and seals also constitute the rich wealth of Archives. Their improper storage results in their further deterioration instead keeping them intact.

Filing flat of registers and portfolios instead of being placed upright is advantageous. Dust does not penetrate into them easily when they are filed flat and, therefore, they are less damaged. The serious disadvantage of the system is that the signatures and cachets (seals impressed on instead of appended to the documents) generally at the bottom of the paper are more easily damaged by rubbing on the shelves. As such, for the preservation of seals, storage upright position of the envelopes containing charts is superior to that of storing flat. It prevents charts from being pressed together.

Map, Plans, Charts and Drawings of larger size need special care in storage. Small maps and plans should preferably be stored flat in ordinary plain filing cabinets but they may also be kept on shelves properly rolled and wrapped in stronger and quality paper.

The wall hanging is the traditionally recognized system of filing maps, charts and plans having large size. Vertical and horizontal cabinet made of wood or steel have got recognition for filing of such materials in modern Archives and depositories.

Bound volumes and books should be shelved upright. Though, weight of the sheets of book or volume shelved upright through pulls away from the spine of the book causing damage to the binding, yet books and volumes so shelved save space and facilitate retrieval. For proper preservation, books should be supported by one another or shelves reasonably filled by book supports or book ends. Large and oversized volumes, however, should be shelved flat.

Irrespective of any storage system, corrosive metal clips, pins and rubber bands, etc. should be replaced by non-rusting material such as fraises, plastic chips, etc. Folded documents and the dog-eared documents from the corners should be flattened. The material infected by mildew, fungus which propagates rapidly, should be segregated, removed and treated as a matter of urgency. Heavily infected materials should be fumigated and followed by treatment with 10% solution of thymol in methylated spirit or methyl bromide. Any small tear in the documents should be mended properly to prevent it from further damage.

Pollution Free Stack Area

The active atmospheric pollutants are available in the outdoor but in Archives and Libraries, making use of modern aids for scientific conservation of documentary wealth, are producing disastrous gases polluting indoor environment of the strong room and act for deterioration of records. Though washing of air in repositories by alkaline solutions is the old system to get rid of outdoor pollutants to some extent, yet a long term programme, keeping in view the whole environmental scenario with a provision of annual revision is the need of hour to maintain a congenial environment in the repository effective to protect the records. To achieve it focus is invited to the following points:

(1) Preventing entry of pollutants.

(2) Elimination of the sources of pollution inside the building.

(3) Deactivation within Archives and Libraries.

(4) Promoting use of enclosures protecting records.

Preventing Entry of Pollutants

The concentration of pollutants inside the repository may be reduced if their entry from outside is prevented. For the purpose, in both types of repositories, air-conditioned and non-conditioned, air-washing system with alkaline solutions which are effective to dissolve acids, salts and other particulate in the air should be introduced. Besides, activated charcoal, carbon absorbing filters which have also been found useful for removing gases and vaporous may be employed. Modern filtration and absorption units, both mobile or stationery may be made part of the Archives and Libraries for the purpose.

The knowledge of material such as cellulose nitrate, acetate foil and collodion prints etc. which have been identified as the sources of pollution in Archives and Libraries, helps a lot in planning pollution control. Periodical survey of storage area, immediate removal of the materials from the area if found generating harmful gases, are the remedial means for controlling pollution within. Modern investigations have proved that removal of oxides of nitrogen is more difficult than sulphur dioxide.

The awareness about local pollution including waste materials to be disposed off and means of disposal in the area where Archives and libraries are located, play a significant role in controlling pollution. The knowledge of wind direction which varies throughout the day, would help in preventing entry of pollutants by adjusting the direction siting control of ventilation intakes into the buildings. Besides, elimination of lead content in fuels like petroleum oil and coal, forbidding the open burring of rubbish and wood, use of liquefied gas as substitute of alcohol and motor fuels could bring about a reduction in pollution to some extent.

Deactivation or Removal of Pollutants within the Archives

Removal or deactivation is the ultimate defence against acidic gases present in the atmosphere. Materials in making of a building plays a significant role in this respect. Cement, stone, concrete, lime plaster, wood work and furnishing materials are good absorber of sulphur dioxide. As such, the preferential use of these material in buildings may help in controlling pollution. Since some of these materials work as active source for creating dust with the passage of time and therefore, their use is useless unless they are sealed without adversely affecting the absorption character.

Renewable wall hanging paper, textile and chalk laden paper are now considered a very effective source for removing sulphuric acid as cellulose itself can retain good amount (say 50%) of this acid. But use of these materials may increase fire risks. Therefore, they should be made fire proof before use and it can be achieved impregnating the materials with sodium borate (borax).

Zeroxing equipments are identified as the potential source producing ozone. As such their use in the close vicinity of stack areas should be restricted.

Filtering system is a good device of cleaning urban air. Most of its contaminants i.e. Aerosols, noxious gases are minimized resulting in reducing chemical imbrittlement of documentary heritage. Hundred per cent effectiveness in any system of filtering air is pointless to think as contaminated air let into area every time when door or window is opened.

Washing of incoming air is a good device to wash out acidic gases such as sulphur dioxide, carbon dioxide and oxides of nitrogen etc. Alkaline water spray (8.5 to 9.0 PH) is effective to remove 100% of sulphur dioxide gas. 50% of this gas is removed in recirculating of air through activated charcoal filters. Suspended aerosols are screened from air by filters, spun out by cyclone devices or precipitated electrostatically. The system of electrostatic precipitation however, has one disadvantage and that is high discharge potential which catalyze the air forming ozone and oxides of nitrogen hazardous for records.

Use of Enclosures for Protection of Records

Showcases made of unstable materials such as plywood, chip board or wood are generally used in exhibitions for displaying manuscripts and paintings. Lining of showcases other than wood with aluminium foil to make them impermeable to all gases and vapour is safer to protect the displayed material from atmospheric hazards. The wooden material such as boxes etc. cannot be made impermeable to gases as by leakage through gaps or permeation through wood provide access to the environmental gases. The buffer box material, however, is the best material for such boxes but cannot be afforded all the time by all. Acid free chalk laden paper as wrapping material in the circumstances, is the right answer. Acid free buffer docket covers, cloth and carton boxes provides adequate protection against environmental pollution.

Besides, following arrangement in the buildings of Archives and Libraries will be helpful in deactivating the pollutants of various nature:

(i) Close inspection of Archival and Library material known to emit damaging chemicals and their segregation as a matter of urgency, if they are found to do so.

(ii) Green vegetation in the surrounding area of building.

(iii) Isolation of containers and show-cases made of plywood and chipboard in Archives and Libraries.

Preventive Care Against Living Organisms

Maintaining ideal storage conditions in the strong rooms is certainly helpful in discouraging the growth of insects and other microbial organisms in the area. But, practically it is not possible to maintain ideal conditions and to have insect free space all the time. Moreover, sometimes, insects and fungus cells get entry in the Archive and Libraries as foreigner and their presence remains unnoticed unless they attack documentary wealth. Perhaps, it was realized long back by custodial agencies and they made some special arrangement other than mentioned previously to discourage their growth and stay. For the purpose various type of insecticides and fumigants have been used in the past.

The protection of documentary heritage from insect attack is seldom an easy matter. Though some methods of treatment are there which render attackless likely. Books and vols in frequent use are never attacked. The tightly packed records in the shelves are more likely to be attacked than those which have plenty of space. Likewise materials shelved in well eliminated areas are less likely infected by insects of various kinds. For preventive care, various type of insecticides and fumigants are employed as below.

Role of Herbs and Plants as Disinfectant and Insect Repellent

The problem of insects to devour documentary wealth is old one and various herbs and plants have been used as insect repellent and disinfectant. Margosa leaves, tobacco leaves, Ghorabach, dried red chilies and slough of snakes are some such popular materials of this category which were used as preventive remedies against biological agencies of all kinds. The use of natural insecticides and fumigants belonging to herb and plant origin has become more relevant today because of coming into notice the adverse side effect of commercial insecticides and fumigants.

Margosa (Neem) leaves dried in shadow are kept interleaved with textile, in carbon boxes and in other containers containing records. Wet leaves stain the documents, therefore, their use should be avoided. Moreover, leaves tied in thin cloth making small packets are suitable to be kept at various points of stack areas. Small tin-containers with perforated covers may be suitable containers for keeping dried leaves, which may subsequently, be placed in the different part of stack area. About 100 gm. of leaves is sufficient to make a cloth packet or filling in the tin container. After every six month the leaves should be replaced by the new lot. Quarantine a principal constituent of neem leave is responsible to keep insects away from the material.

Tobacco leaves have a very pungent smell and they are being used by Sharma Museum, in Jaipur. This Museum has in its custody a very rich collection of old manuscripts of various kind. Small amount, say 10 grams. of tobacco leaves is sufficient to disinfect properly one m^3 space. It is put inside the glass covered shelves, in the said Museums. A small amount (about 10 gms.) of it may be kept in a small tin container which is used by tobacco chewers in India for keeping *Chunah*." The perforated cover of container will provide passage to tobacco fumes diffusing in the area. Such containers filled with appropriate amount of tobacco should be kept at various point of stack area as per requirement. Curator Sharma Museum, informed that the effect of tobacco preventing insects etc. is very encouraging. To maintain the concentration of tobacco fume in the closed area, the old lot should be replaced after 6 month.

The same technique may be applied for placement of dried red-chilies in the storage area. Investigation about the placement of remaining herbal based insecticides and fumigants is in progress in the laboratory.

Use of Commercial Insecticides and Insect Repellents

The use of commercial insecticides, insect repellents and disinfectant for preservation of records is very common today. Some of them are used in solid or crystal form, some as spraying agent while some as surface coater.

Naphthalene Bricks

Naphthalene ($C_{10}H_8$) is a time tested commercial insect repellent. It is combustible, white crystalline, volatile organic material derived from coaltar and used in the form of bricks of special size say 5" x 2.5" x 2" (L x W x H). The size is convenient to give a surface area for evaporation of naphthalene forming proper fume concentration irritating insects in the area. M.L. Roonwal and P.N. Chatterjee in their studies on controlling bookworms suggest the air heating of infected documentary materials in suitable ovens at 60°C for 4 hours and at 70°C for 3 hours. Further they suggest, the exposure of infected material to sun for 5-4 hours in Indian Plains where high temperature prevail in summer, to remove insects from records and books.

The frequent exposure of infected records of all kind in sun in general and treating them at elevated temperatures may effect them adversely and decayed, old records made of ground wood pulp particularly.

The bricks of specific size (mentioned above) are kept at a distance of two meters in every dimension. Since the fume formation is affected by various parameters such as heat, light and movement of air which varies from place to place and region, to region, therefore, we cannot stick to this distance in every season and every place. In a region where atmospheric temperature normally ranges between 23°-30°C, the suggested gap of two meters between the two bricks is suitable while at temperature below and above, this distance may be decreased or increased 1 meter to 2.5 meters respectively depending on the conditions.

Spray of Insecticides

Climatic condition of tropical and sub-tropical regions is favourable for insect and fungus growth and their propagation. As such, the disinfection of stack areas from time to time by suitable insecticides in general and in transition periods in particular helps in curbing cells of various species of biological origin and preventing their growth. Though numerous insecticides are available for disinfection but organizations like Archives, Libraries and Museums where people work, only time tested materials which are neither dangerous for people nor documents are suggested to use. Baygon, clenofix, (contents of this insecticides are Pyrethrum, D.D.T., Pine-Oil), Shelfox, (contents are Dieldrin, Pyrethrum emulsion) and Pip (contents are, Pyrethrum and D.D.T.) come in this category, D.D.T. once upon a time was used for this purpose but recent discoveries have found it carcinogenic and, therefore, it has been banned for further use.

These insecticides for disinfection of areas are sprayed on the walls and floor but not directly over the record material as they may stain the paper and may increase the rate of deterioration.

Fog of pip or any other suitable insecticide works more effectively than the spray particles in annihilating insects and micro organisms. Since fog is capable to enter the inter-spices of books, volumes and files, can eradicate insects hidden therein. For preparing fog, a suitable machine known as Swing Fog Machine is used.

For disinfection of areas, the day followed by a holiday is very suitable. Before spraying or fogging all the windows, ventilators, and other opening of the area are closed to keep the fumes captive for effective results.

Coating Materials

For protection of documentary materials (specially book covers), various commercial insecticides have been directly used over them as surface coater. A.W. Mckenny Hughes, had recommended the coating of Ammonium and Potassium Thiocyanate on paper or covers to protect them against inspects. He has mentioned that this insecticide provides high degree of protection to records against vermin. Besides, 5% solution of mercuric chloride was also used but being very poisonous and persistent hazardous to human being and contributing acidity to the materials applied, it is not recommended for use.

Brushing or spraying of potassium antimony tarrate (Tartar Emeric) in water and 1-2% solution of tricresyl-phosphate in refined kerosene over the documents protects them against silverfish.

Coating of DDT, Pentachlorophenol, Pyrethrum/Piperonol, Aldrin, Gammaxane and BMC in urea fermaldeyde resins and butoxide have also shown promising results in protecting documents against insects. But application of resins is not suitable over leather as it entirely alters the appearance, looking shiny, after treatment.

Spraying or coating of residual insecticides like BMC, Chlordane in Kerosene or their use as emulsion have also been tested with positive results.

Sulph Arsenic

Prof. K. Sitaram Iyre of College of Science, Trivendrum, had much talked about the use of Sulph-Arsenic as insecticide for the preservation of documentary heritage as surface coating agent, in a meeting held at Bangalore on 20th March 1935. He recommended the application of arsenical insecticide solution, prepared by mixing one lb of arsenious oxide and ½lb. of sodium sulphide in 10 gallons of water, on books, manuscripts and documents. In bound books and volumes the solution was applied between the cover and fly leaf since these parts are more succinctly attacked by insects. The effect of solution once applied was to last for five years. The treatment cost in 1946, was nominal say five *annas* (thirty *paisas*) 70,000 volumes of medium size.

As the time went by, the information about the use of sulph arsenic was picked up by various libraries of the country. Natonal Archives, a leading concern in document preservation received queries from various organisations about the suitability of the material for protection of books and other documentary heritage. Consequent upon, S. Chakravorti and P.C. Majumdar of the Deptt, took up the work for experimental study in 1947 to know the effect of coating mixture on durability and permanency of paper. Finding of the study revealed the fact that the solution is really effective to prevent the books from insects but durability of paper is adversely affected. Application of sulph-arsenic solution brought about a substantial reduction in folding endurance by 32.72% tensile strength by 14.14% and bursting strength by 18.17% when documents were subjected for accelerated aging (at 100 ± 2°C for 72 hours). Test data are available at Appendix IV.

The reason of adverse effect brought about by the solution to paper, is perhaps because sulphur in sodium sulphide becomes free in the colloidal form and get oxidized gradually into sulphur dioxide and its acids. The deleterious effect of the sulphur dioxide gas and its acids is well known.

Viewing over the adverse effect brought out by the solution, its use could not be recommended inspite of its low treatment cost. In the meanwhile, it was also expressed by the study makers that the solution applied between the cover and the fly leaf would not involve the decay of other part of the book/volumes and would prevent insect attack. But if the cover and the flyleaf deteriorate, the binding would certainly give way much sooner and the advantages would be offset by the sharp increase in rebinding charges.

Gammexane ($C_6H_6Cl_6$)

Once upon a time gammexane because of its toxicity and lasting effect against insects and cold-blooded creatures, was being used to control them in Archives and libraries. A practical study conducted by S. Chakravorti, of N.A.I., revealed the fact that mere exposure of paper to gammexane vapours, brings about substantial decrease in its tensile strength. Shri Chakravorti, mentioned in his report that the damage to paper in fact is much more greater than what we would expect from its accelerated aging. It also adversely affects the iron gall inks. The chemical is banned today for its carcinogenic character.

An Ointment with Cederwood Oil

E.A. Buck in his article 'Bookworms' has suggested to store insect infected books and documents for protection against, insects in earthen vessels, after they are ointed with cedar oil. It is very old technique perhaps prevalent near about 3rd century B.C.

Preventive and Curative Measures Against Rodents

The records are destroyed knawing by rodents normally because of revengeful attitude, interest of food and sometimes as notorious activity. For planning effective preventive strategies against the animal following measures are adopted:

(1) Like other insects, rodents are also attracted by the food items, therefore, entry of edibles in the stack area, should strictly be prohibited.

(2) No granary or store house should be allowed to function nearby the Archives and Library buildings and in their campus.

(3) All the holes, cracks and crivices in the repository as well as in the other part of buildings should be covered with concrete and cement as a matter of urgency.

(4) Outlets in general in the Archival and Library buildings, particularly of repositories, should be covered with metal gauzes. Pipes reaching different floors of the building should also be covered with the gauzes from both ends.

(5) Push-pull type of doors flushed with ground in the repository should be fitted so that they may automatically shut after use preventing entry of animal from other parts of building.

(6) Windows fitted with metal grills should be followed by wire mesh fittings.

If rodents, inspite of above measures enter the repository, there is no way out but to kill or trap them. Various baits, poisons like zinc phosphide, warfarin are used to kill them. The use of trappers also give the right answer in removing the animal from the repository or any closed space. While annihilating the animal, it should be ensured that there is no point in the closed area for their escape and that can be achieved by executing preventive measure strictly. A failure in executing plan with full care may result in loss of material due to knawing by the animal as an act of vengeance.

CURATIVE CONSERVATION

Curative conservation activities come on the play when documentary wealth is attacked by vermins and fungi. The purpose of curative conservation is to destroy the hazardous organisms set on the records without bringing about any adverse change and damaging wealth.

Today various fumigants, insecticides, pesticides, and poisons mostly commercial are being employed to get rid of the biological agent. But early steps, taken to destroy them in Indian sub--continent were based on natural treatment. I cannot do better than quoting Purendu Basu, who has mentioned the exposure of affected books and volumes to the sun in his article, "The Enemies of Records" as remedy from agencies. But at the same time he talks that this may be effective against vermins and fungus but discolours and ruins paper. Obviously, the remedy is as dangerous as ailment. Covering/wrapping materials like cloth, board, wood and containers are still exposed to sun to destroy any insects but the practice of exposing documentary wealth of organic nature has completely been stopped. The reason behind is the destructive nature of sun to the record materials.

Fumigation of insect and fungus infected records, has emerged today as the major technique of curative conservation.

It seems that the process of fumigation using commercial fumigants and lethal gases got the way in finding out suitable means and ways to sort out problem faced by Huntington Library caused by book-worms or more specifically, the Sitodrepa Panicea. The problem was so serious that library authorities when failed to find out suitable method of their own to combat the insects, they sought advice from experienced libraries of the world.

Out of various suggestions received by Library, one was related to sprinkling of a little fine pepper on the shelves containing books. Another institute advised the use of ether but failed to say whether it should be administered to the book-worms or the librarian. One librarian from the tropics suggested the use of flit but at the same time mentioned that this is not sufficient. Research and development activities relating to annihilation of vermins in the tropics and sub-tropics had been very nominal in the past as no any serious problem was faced. The reply received from Monsignor Tisserant, Proprepetto of the Vatican Library, for combating the problem, was however, very meaningful. It wrote, "We have great trouble with book-worms. The damage have been greatest since 1912, at the time, the books were taken from boxes and shelved in stacks. We used against them carbon disulfide in a zinc box. So far as the effect, it mentioned, the process was slow as the box was small and books to be disinfected were many. Further, it suggested the use of cyanic acid gas against a caterpillar, a type of insect damaging leather bindings. The results of treatment were good. But Vatican Library could not conduct more experiments of the like because the firm they engaged for the fumigation work disappeared in the mean time.

Obviously, the use of carbon disulfide and zinc box, was the first attempt of its kind in the history of fumigation of documentary material in 1912. Thomas M. Iams, in his article, "Preservation of Rare Books and manuscripts in the Huntington Library" has stated that in 1917, the United States Bureau of Entomology reported the fumigation of libraries with hydrocyanic gas but primarily for the destruction of external feeding pests as silverfish and cockroaches. Huntington Library people by now had realized that the problem which they were confronting with, was a serious one and if they want to sort out it permanently, they would have to take strenuous measures. As such, soon they instigated a series of experiments, of treating the infected books and records with hydrocyanic acid gas (HCN), carbon disulfide (CS_2), Carbon tetrachloride (CCl_4) and ethylene chloride or dichloride $C_2H_4Cl_2$, a newly appeared fumigant in the market at that time. Since, it was not sure that the gases and fumigants they are experimenting with have no deleterious effect on the record material, to continue with the experimental work therefore could not be possible. Now, they turned their attention towards vacuum fumigation. Dr. Hagen's experiment of putting rarities under the glass bell of an air pump and drawing out air of it, had broken the ice for vacuum fumigation. Experimenting with this process it was found that technique almost allows perfect penetration of gases or fumes without opening every book so as to expose as great a surface as possible to the fumigant. Commending the technique, Thomas M. Iams has said, "We had at last found an excellent method of fumigation but still searching for an ideal fumigant".

The mixture of ethylene oxide and carbon dioxide, in a liquid neither, inflammable nor explosive being marketed at that time under the trade name 'Carboxide' was accepted as ideal fumigant. The result of the mixture fumigant was very effective and the battle on fighting book-worms was won by the Huntington Library. The efforts of library, against the injurious pest, paved the way for the use of commercial fumigants and vacuum fumigation techniques in the field of Archivo-Library Conservation.

Fumigants for Sterilization of Records

There are various techniques and method for sterilization of records but fumigation in general and vacuum fumigation, in particular, is the most effective processes. Each process of fumigation requires a fumigant and suitable equipment. Some fumigants are used alone while some as mixture. Fumigants are available in solid, liquid and gaseous states. A table of fumigants tried in Archives and Libraries alongwith their required quantity is given below.

Table 4.1 : Showing Fumigants and their Appropriate Amount for Effective Sterilization of Records

(A)	**Fumigants which are used alone**	**Amount required per 1000 ft^3/M^3**
(i)	Para-dichloro benzene	1.5kg/M^3
(ii)	Formaldehyde	–
(iii)	Methylbromide	2lb/1000ft^3
(iv)	Thymol	120 gm/M^3
(v)	Ethylene oxide	2lbs/1000ft^3
(vi)	Naphthalene	1/2lb/M^3
(vii)	Carbon dioxide	20lbs/300ft^3
(viii)	Nitrogen	
(B)	**Mixture Fumigants**	
(i)	Killoptera (Mixture of Ethylnedichloride and Carbon tetra-chloride, 3:1)	14lbs/1000ft^3
(ii)	Carbon-Disulphide and Carbon tetra-chloride (1:4)	6lbs/1000ft^3
(iii)	Carboxide (Mixture of Carbondioxide and Ethyleneoxide (9:1)	30lbs/1000ft^3
(iv)	Methyl formate and Carbon dioxide (2:5)	28lbs/1000ft^3
(v)	Ethyle oxide and air (1:1)	

Required Characteristic of a Fumigant

A good fumigant should be characterized by adequate volatility to produce toxic concentration in an enclosed space killing vermins. Some fumigants are volatile at ordinary room temperature (22°C), while some require additional heat to produce fumes. Fumes of some fumigants are heavier than air, tending to sink downside.

The physical state, ambient temperature and the quantity of a fumigant to be used for fumigation require attention while they are selected for the purpose of fumigation. The rate of diffusion depends upon the molecular weight of gas, absorptive capacity of material and size of the opening through which the gas has to pass. To ensure uniform distribution such fumigants are often applied through nozzles from the top of the space of the equipment in which fumigation is conducted. The fumigants in solid and liquid state which are not volatile at ordinary temperature require heating to produce sufficient concentration of fumes effective for sterilization. Thymol and Naphthalene are the fumigants of this category. However, the fumigants like tetrachloride, carbon disulfide and ethylene chloride (all in liquid state) and para-dichloro benzene (solid state) are volatile at ordinary temperature and need not heating.

The quantity of fumigant is decided keeping in view the temperature, length of exposure, the amount of absorption and adsorption by the material to be fumigated and the air-tightness of the vault in which fumigation is carried out. The quantity required of each fumigant for the purpose of fumigation has already been indicated against each of the fumigant.

The Role of Temperature in Fumigation

The effective fumigation means the maximum, (ideally total) sterilization of insects. Temperature is one of the important factor to achieve it. Insects are generally found dormant below temperature 10°C and at this state, it is difficult to kill them with fumigants. For proper action of fumigants, the toxic action should be effective and that depends upon the penetration of fumes through respiratory system, the activity of insects and, susceptibility. Near about 24°C, the insects are almost highly active. As such, there is no need for exceeding temperature within the enclosed space of fumigation chamber or vault for the purpose, in tropical and sub-tropical regions as this temperature range in the regions almost rests in summer. However, if fumigation is to be conducted in the regions or in the areas where normal temperature is below 20°C, the suitable fumigation technique would be that which makes use of thermal system for enhancing temperature inside the chamber or vault. Selection of a suitable fumigant viewing over the nature of material should be carefully made as wrong selection may adversely affect the record. For example methyl bromide affects the leather and photographs.

Role of Carbon-dioxide in Fumigation

The role of carbon dioxide in fumigation is two-fold. First, it reduces the inflammability of fumigants and secondly, accelerates the toxic action of fumigants by inducing the openings of the breathing pores of insects.

EQUIPMENT AND TECHNIQUE OF FUMIGATION

Vacuum Fumigation

The technique of fumigation is recognized as the most effective and speedy means of fumigation. Archivo-Library world is indebted to Dr. H.A. Hagen of the Department of entomology of Harvard University, for developing a modern fumigation technique destroying book-worms and sorting out the problem faced by Huntington Library. At last, out of series of experiments, Hagen, conducted an experiment, putting rarities under the glass bell of an air pump and drawn out the air. After an hour, whatever he observed that was encouraging for him that all the larvae were found dead. He communicated his results to the Huntington Library puzzling with the problem of book-worms. The eggs, chrysalis and beetles are harder to destroy. To ensure full success, Huntington Library people involved the use of fumigant with the vacuum treatment. The earliest vacuum equipment was got designed by the Authorities of Huntington Library, for sterilization of bookworms infected records. The insect by no other means but by vacuum fumigation technique only could be exterminated and destroyed. The dimension of the fumigator designed was five feet in diameter and sufficiently large to accommodate five or six library trucks full of books at a time. The actual operation of the apparatus was very simple and did not need the services of a trained engineer or a fumigant expert, says Thomas M. Iams.

The vacuum Fumigation Equipment owned by National Archives of India was acquired in 1951. The equipment is fitted with evacuation system, mixing tank and system of disposal of unused fumes of a fumigant. Besides, pressure and temperature measuring equipments also get place in the chamber.

Inceptionally, 'Carboxide' (a mixture of ethyl oxide and carbon dioxide in ratio 1:9) was used as fumigant. 30lbs. of the fumigant is sufficient to produce fumes of requisite concentration within the chamber to kill insects within a time duration of 5-6 hours. Nothing is exaggeration to say that it is a complete fumigation technique, as it not only destroys the insects but their eggs as well. The significant characteristic of the technique is that a lot of records, approximately 1800 books at a time may be fumigated within a few hours. As such, it is also termed as mass fumigation technique.

M.N. Sarkar, National Library, Kolkata, has mentioned that for the destruction of fungus, however, the exposure time and temperature used to be higher than that required to kill insects and their eggs.

The important steps involved for conducting fumigation in vacuum fumigation technique are:

(i) Cleaning of chamber for dust and other foreign materials.
(ii) Loading of material to be fumigated.
(iii) Closing of door and fixing it properly ensuring complete tightness.
(iv) Mixing of Carboxide components in a mixing tank. Since the ethylene oxide is declared carcinogenic and banned, the carbon dioxide alone is used now at fumigant in National Archives of India. It is pumped into the chamber to a pressure of 7 atmospheres to kill insects. Nitrogen gas could also be used as substitute of carbon dioxide.
(v) Creation of vacuum in the chamber. It is achieved by a suitable system through suction arrangement fitted with the chamber.
(vi) Injecting of suitable fumigant.
(vii) Disposal of unused fumes of the chamber after 5-6 hours, when cycle of fumigation completes. The disposal system should be effective, leak-proof so that unused fumes/ gases could be pumped out of the room to the open environment.
(viii) Injecting of atmospheric air equating external atmospheric pressure.
(ix) Removal of fumigated material.

The vacuum equipment of different capacity are manufactured by W.J. Alcock & Comp. Pvt. Limited, 7A, K.S. Road, Kolkata in India. Thermostat fumigation vaults of different sizes are manufactured by M/S Narang Scientific Works, New Delhi, under brand name Caltan. Fumigation equipment of both the types are being used by National Archives of India.

Fumigation with Methyl Bromide

Methyl bromide is extensively used for fumigation of food-grains in granaries and can be effectively used for sterilization of records and books, in Archives and Libraries.

It is colourless, transparent, liquefied gas of volatile liquid of low boiling point say 35°C. The gas is three times heavier than air. The gas being odourless requires addition of 2% chloropicrin (Teargas) by weight for identification of its presence. The presence of gas, however, may be detected by using halide detector lamp. The flame of spirit lamp which is invisible normally turns green light to dark blue depending on the concentration of methyl bromide in the atmosphere.

The chemical is soled and supplied in steel cylinders. Cans containing 1lb of methyl bromide are also available which are suitable for small scale fumigation. An amount of 205 gms. is sufficient to produce required concentration in one cubic meter of space for killing insects. The fumigation can be carried out at normal temperature of 25°C. The cycle of fumigation completes in 24 hours.

Gas-tight covers, vault and vacuum chamber are the suitable equipments for carrying out fumigation.

Fumigation in Gas-Tight Covers

Gas-tight covers made of rubber proofed double texture fabric with aluminium finish are suitable for carrying out fumigation in the open. Such covers are manufactured by Central Ordinance Depot, Kanpur. The two convenient sizes of covers are 1000 and 2000 ft. One cover can be used 100-125 times for fumigation.

Process of Fumigation

The cover is draped over the material to be fumigated in the stack and its edges all around lying flat on the floor are sealed with mud mortar.

The cylinder containing (fumigant) is placed over a spring balance, kept outside the cover. The valve of cylinder is connected with one end of rubber tude and the free end of it is connected to a 'T' shaped glass/copper tube attached with the cover on its top. Two or more branches of tubes may be laid ensuring uniform distribution of the fumigant in the cover.

The requisite amount of fumigant is now released and the valve of cylinder is closed. The outer end of the delivery tube is closed with stop cock and disconnected from the gas cylinder. Gas leaks are checked using halide detector lamp and immediate action should be taken in case leak is detected.

Opening of Cover

After the fumigation treatment is over, the sealing mud is removed and on the leeward side is lifted and raised to the top of the stack. Since fumigant is poisonous, the certain precautions are must to be taken care of while opening the cover:

(1) No one should be allowed within 10 meters on the leeward side of the stack.

(2) The work in the fumigated stack should be started after an hour of opening the cover.

VAULT FUMIGATION

A well plastered, painted and air-tight room provisioned with well gasketed fitting doors/windows operable from outside, with adequate ventilation facilities is a suitable vault for carrying out fumigation. Ensuring effective treatment, depending on the size of vault one or two exhaust fans should be got fitted for quick degassing. Since most of the fumigants are heavier than air, the windows and exhaust fans should be located near the floor level. Besides, the room should be tiled and the crevices between the door/windows leaves alongwith their frames should be sealed with mud mortar or gummed paper during fumigation in order to check absorption of fumigant. In case the vault is set up inside the building, the ventilating stack must extend exterior of the building preferably above the roof. One such vault is operative in CRDC, College of Agricultural Banking, RBI, Pune. It was constructed on the expert guidance of NAI.

Technique

The material to be fumigated is placed in suitable racks in the vault after cleaning. One end of the rubber tube is joined with the valve of cylinder containing fumigant and the free-end of it is, subsequently joined to a 'Y' shaped glass or copper tube. For effective fumigation proper dispersion of fume in the chamber is a must and that can be ensured by stirring the inside atmosphere of the vault. Vault is now closed and sealed. Before sealing, however, it should be ensured that windows are properly closed and sealed.

The adequate amount of fumigant is injected and the cylinder valve is closed. The outer end of the delivery tube is closed with stopcock and disconnected from the gas cylinder.

After cycle of fumigation is complete, the doors and windows on the leeward side are opened. In case, exhaust fans are provided in the room, no one should be allowed in front of the exhaust fans within a distance of 10 meters for an hour. Workers are allowed to enter the chamber after about 2 hours of opening of the windows and doors.

Equipment for Fumigation by Para-Dichloro Benzene or Kiloptera

An air-tight chamber of adequate capacity is required to carry out fumigation using paradichloro benzene or killoptera. A Godrej type almirah, if converted air-tight and shelves are perforated and made adjustable, may be suitable equipment for the purpose. Such chambers are being used in the various repositories of India, including National Archives.

Materials to be fumigated are placed or hanged suitably over the shelves ensuring no further damage due to any reason during the process of fumigation. The books and volumes with hard covers, however, can be put in inverted 'V' Shape as shown in figure 4.8. The files with thin covers, however, incapable to carry the weight of documents cannot be placed on the pattern suggested above. For the purpose, special stands, which can be accommodated in the shelves, are used for hanging of files and loose sheets.

Para dichloro benzene crystals, 1.5 kgs. or 225 ml. of killoptera is suffice to produce adequate concentration of fumes per cubic meter at ordinary temperature for killing insects. The vapours of the fumigant are heavier than air. As such, suitable amount of either of the fumigants is placed on the upper-most shelf and chamber is closed. If chamber is not air-tight that can be converted by pasting adhesive tape on all the opening suspected leakage.

A time duration of 7-8 days is required for completing the process of fumigation in case of para dichloro benzene while 24 to 26 hours if killoptera is used. This fumigation treatment will annihilate insects and their larvae leaving eggs unaffected. Normally, eggs after 20-21 days from the date they are laid down, become mature to larvae and, therefore, for complete sterilization, refumigation of records after expiry of this period is necessary.

Naphthalene Fumigation

Naphthalene a white crystalline, volatile chemical, which is used as insect repellent in Archives and Libraries is well known chemical. Though, it is considered, a week insecticide because of its low vapour concentration at ordinary room temperature, yet it works as an effective fumigant at temperature range of

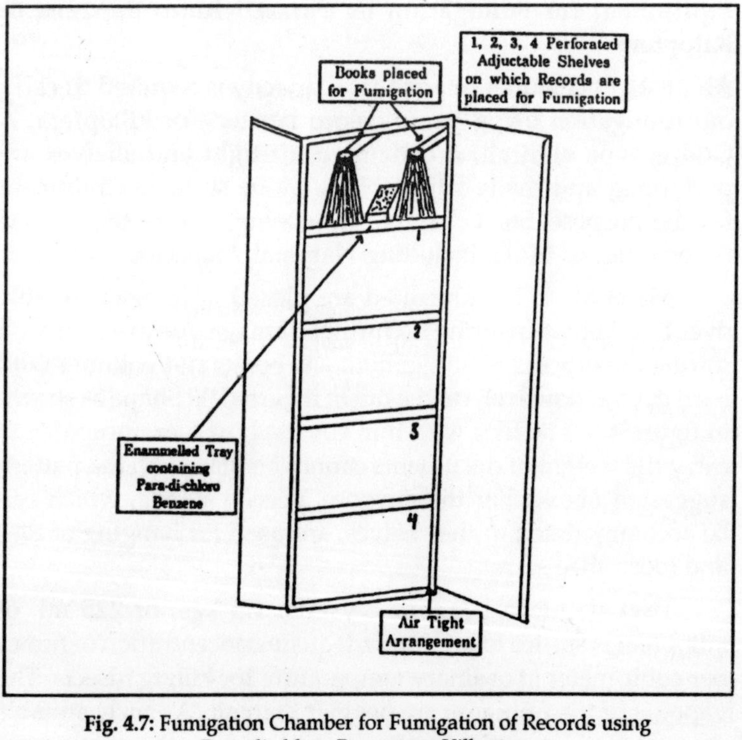

Fig. 4.7: Fumigation Chamber for Fumigation of Records using Para-di-chloro Benzene or Killoptera

30-35°C. It has been observed that naphthalene in molecular form is much more effective as fumigant than carbon disulfide (CS_2), Para-Dichloro benzene and carbon tetra chloride.

It is evident from the data of an experiment conducted in the research Laboratory of National Archives of India available at appendix V. A required concentration for paralyzing insects like cockroach, silverfish and larvae of *Gastrallus indicus* is achieved by 1.50 lbs. of fumigant per cubic meter at a temperature range of 30-35°C in 1.30 hours. This concentration is achieved by passing a hot air blast through a vertical column of Naphthalene powder injected into the chamber.

It is the most suitable fumigant for tropical and sub-tropical regions where temperature range normally rests above 25°C. Since sterilization can be done without any additional heat supply to the system, the process will prove cheaper.

Though, test data are silent about its effect on paper but it may be presumed that it does not adversely works.

Insecticides Controlling Insect Infestation

S.M. Nair has suggested the use of following insecticides for controlling insect infestation effectively:

Table 4.2 : Insects and Insecticides

Insect		*Insecticides*
(1) Book-lice	(i)	P.B.C. or Para dichloro benzene as repellents.
	(ii)	Spray of pyrethrum.
(2) Book-worm	(i)	Fumigation with Methyl[+1] or Ethyl Bromide or Carbon-di-sulphide.
	(ii)	Exposure to vapour of P.B.C.
(3) Cockroaches	(i)	Spray or sprinkling dust of 5% DDT[+2], BHC, Pyrethrum.
	(ii)	Sodium fluoride and Gypsum (1:1) as poison powder.
	(iii)	PBC as a repellent.
(4) Silverfish	(i)	Spray or sprinkling of dusts of 5% DDT, BHC. Or Pyrethrum Mixture of Para dichloro benzene, benzene[+3] and creosote (1:1:1) as repellent.

+1 +2 +3 - Banned now

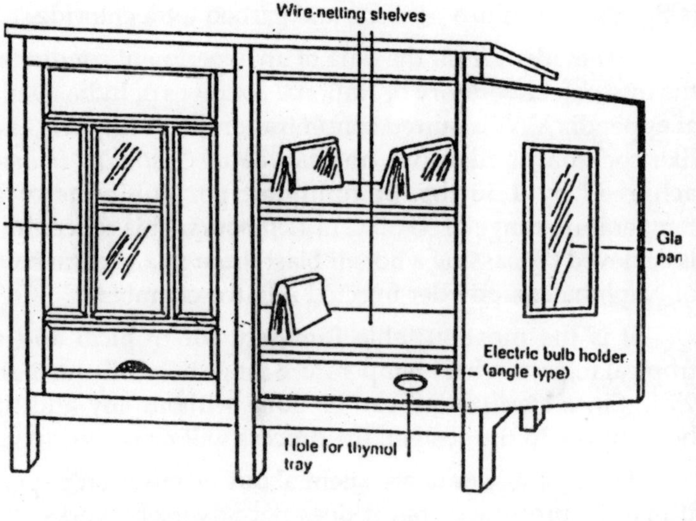

Fig. 4.8: Thymol Fumigation Chamber

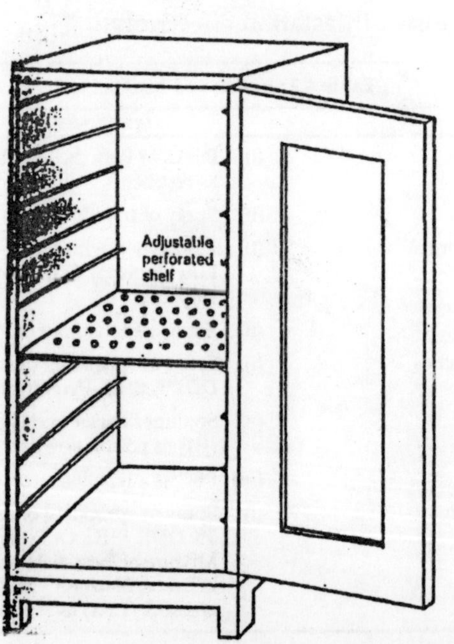

Fig. 4.9: Single Door Air Tight Steel Fumigation Chamber

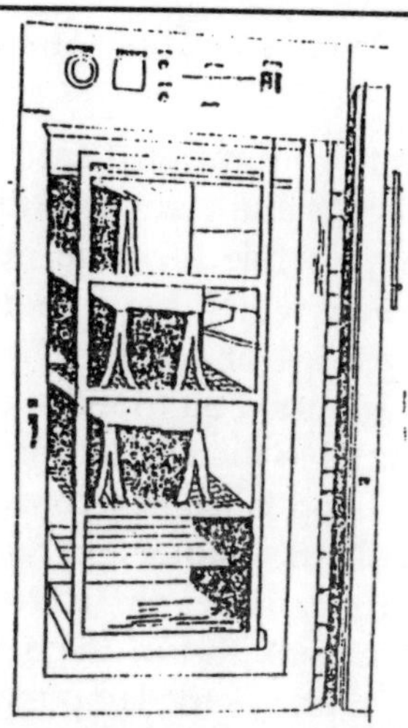

Fig. 4.10 : Fumigation Chamber (Thermostat)

Eradication of Fungus

Fumigation is the effective device for uprooting fungi from the fungus infected records. The fumigants which are normally used for annihilation of insects, are not effective for eradication of fungus. Thymol, is, however, an effective fumigant for this purpose.

An air-tight cup board or box, provisioned with the frame-work of wire net and a suitable heating arrangement in the bottom self is the suitable equipment for carrying out thymol fumigation. Thymol does not give vapours at ordinary temperature and requires heating. A suitable bulb holder or thermostat system fitted on the bottom of the chamber, is the suitable heating arrangement. Besides, the fumigation chamber which is used for fumigation of records using para dichloro benzene or killoptera may also be used for thymol fumigation, if a suitable heating system fitted on the non-conducting materials is arranged in its bottom shelf. The chamber will function as three-in-one i.e. for fumigation of both insect as well as fungus infected records and for their deacidification.

An amount of 120 gms. of thymol gives appropriate concentration of fumes at temperature approximately 40°C for uprooting the fungi. A time duration of 6-10 days is sufficient to complete the cycle of fumigation. The vapour of thymol are lighter than air.

The materials to be fumigated are placed over the frame work of wire net. Books and volumes with hard covers are put in inverted 'V'-shape while files and other manuscripts are kept in tied bundles, which have been loosened properly facilitating free passage of fumes. The suitable amount of thymol in a dish supported on tripod stand is placed over the electric lamp or suitable thermostat for heating. Selection of wattage of electric lamp is made according to the season and the climate. Normally in summer and rainy season, 40-60 watt electric lamp will radiate sufficient heat stimulating vapourization of thymol. But in the areas where climate is cooler and in winter, an electric lamp more than 100 watt will be necessary for this purpose. Every day, to produce required concentration of fumes in the chamber, lamp (electric) fitted in the vault is made on for 3-4 hours depending on the climate conditions. After expiry of fumigation cycle, the chamber is opened and material is exposed to the environment. Subsequently, the fungus is removed using cotton-wool followed by treating the infected places with 10% solution of thymol in methylated spirit. Books and files so treated are interleaved with thymolysised paper. It will inhibit the re-growth of fungus. The treated material is kept under close examination and any re-growth of fungus attack is treated with thymol solution immediately. After few months the records are restored in the original places.

Y.P. Kathpalia, had suggested the use of a mixture of ethylene oxide and air in the ratio 1:1 for treatment of fungus infected records in vacuum fumigation vault. But, ethylene being carcinogenic is not suggested to use as fumigant.

Microwaves, laser and deep freezer treatment are the some latest techniques for sterilization of insects and fungus infected records.

Microwave Treatment

Y.P. Kathpalia, in his article, "Sterilization of Records against fungus, insects and rodents" has suggested the exposure of infested documentary materials for a duration of 1 to 1.30 hours in microwave oven at 60°C to 63°C, for annihilation of insects and uprooting of fungus without adversely affecting the keeping qualities of paper. So treated records are conditioned under normal temperature and relative humidity prior to cleaning and storage. The technique is promising it is said. Unfortunately, no test data has been published so far. However, it is young field for research oriented conservation scientists.

Laser Treatment Technique

Kathpalia and his colleagues working on the technique went by the positive results and hoped that the relevant data relating to the experiment will be released soon.

The laser beam of moderate potency is effective to remove pencil marks without adversely affecting the paper carrier. Keeping this fact in view, they experimented using laser for annihilating insects and uprooting fungus and prompted by positive results. They had mentioned that even insect infection can effectively be countered using laser. Unfortunately, the data of experiment could not be published. It is the young field for researchers in the field of conservation of wealth in question.

The use of Radioactive or gamma rays technique is increasing day by day for sterilization of insects and pests.

Ultrasonic sound waves which are capable to bring about chemical and biological changes have been experimented to destroy bacteria, repelling and killing small animals like rodents.

Talking about the characteristics of fungi, S.M. Niar states that they vary enormously in their response to fungicide. Some type of fungi can tolerate surpassingly high concentration of chemical while some are inhibited by quite low concentration of that. For instance, even very low concentration of thymol and saffron can inhibit the growth of a variety of fungus, Aspergillus niger, whereas they have little effect on certain penicillin species. Pencillium is seriously affected by Para-nitro-phenol and zinc sillico fluoride while Aspergillus remains unaffected against it. These facts are indicative that in controlling fungi, the identification of its species is a must so that the use of effective fungicide may be possible eradicating fungus easily and effectively.

Based on the results of his experiments, Mr. Nair, suggested use of following fungicide:

	Fungicide	Concentration
(1)	Thymol and sodium fluoride	Above 5%
(2)	Mercuric chloride	A very low concentration 0.2%
(3)	Paranitrophenol	3% or above to 5%
(4)	Pentachlorophenol	Very low concentration 0.25%
(5)	Sodium Salicylate and Sodium Pentachloro phenate	10% or above

Specific Consideration in Use of Biocides

The fumigants are poisonous not only to the insects but also to the human beings, therefore, the use of such chemicals should be undertaken with extreme care. Toxicity prevails by injection, inhalation and skin absorption and to eschew it use of masks and gloves should invariably be made. Most of the treatments specified in the chapter should be undertaken after isolating the books and documents from the stacks and reading areas. Files and books, in frequent use should best be treated with volatile chemicals which do not leave any residue after the biocidal treatment is over.

Foxing

Foxing causes widespread damage of books, manuscripts, archival records, drawings and engravings etc. The edges of the books, sheets injecting from the book block and cardboard book binding are the common places where foxing marks begin to develop with passage of time. Besides, they are seen on the objects of cotton material which are exposed for a long period of time. These marks are identified as coloured stains varying from lemon yellow to dark brown. More ordinarily the agency has been named so because of its fox colour.

The foxing marks, have been observed on various kind of paper from 16th century A.D. The paper affected by foxing stains is weaker, more acidic and friable than the unaffected and uncoloured one, suggest studies of the files. The stereo-binocular microscopic examination of foxing stains look like as if they were slightly pressed into the surface of the substratum.

The most light coloured stains having a citric shade are found on low grade paper like news print while dark are characteristic of high quality paper.

The growth of foxing marks is influenced by the technology adopted for paper manufacture; grinding in stead of beating the pulp, dust and pollutants, long exposure in light and moisture fluctuation (temperature and humidity. The stains show intensive luminescence in U.V. light at the initial phase of formation when they are practically unseen in visible light.

So far as the history of research in the nature of foxing dates back to 1930s. Two thoughts on the formation of the marks are prevalent today. One believes that their formation is the result of oxidation of iron present in the paper or the material affected with the agency while another accounts their formation because of development of micro-organisms. However, the recent experimental studies in the nature of foxing have thrown light that the area of paper attacked by foxing is always more acidic.

Preventive and Curative Care

The foxing growth on drawings and engraving can be prevented if they are mounted on acid free cardboard of high quality. Besides, following measures are effective to curb, further development of foxing on the suspected documents/ objects (U.V. light detection).

(i) Washing of affected documents indistilled water or water alcohol mixture.

(ii) Impregnating the affected documents in weak solution of gelatine. The high butter capacity of gelatin retards the rate of acidification of treated documents.

Planning Modality for Curative Conservation

It is clear that adequate know how for countering agencies of biological origin in Archives and Libraries is available in the country. For errorless and effective use, meticulous planning to carry out treatment/sterilization is essential. The practice in various big and small Archival and Library organizations is that they fumigate the records in piecemeal for extermination of insects and uprooting fungus. The material after treatment is restored in the stacks and subsequently, other lot from the same stack is received for fumigation. The process is continued till the last trace of the lot is fumigated. The question is whether this system of fumigation is really effective for complete annihilation of insects and to make records free from them? Answer is certainly negative. The question is why? To find out answer, the following facts are well to be considered.

The insects once set in the repository do not remain confined to the specific area of the stack. As the time goes by, they grow, multiply and spread according to their convenience and choice in the area of the stack and other parts of building. Moreover, some insects make their shelter in the holes and cracks on the walls and ceiling of the building. If one lot of the records and books has been fumigated effectively and restored in its original place, it is not guaranteed that insects from other parts of stack will not reach the fumigated lot. Certainly, after sometime, the insects from the same stack area active on devouring the books and manuscripts will make their way and re-attack the fumigated lot.

Other important fact is that records are fumigated but materials used for storage like cloth, carton boxes, ply wood boards and wooden shelves are seldom given anti-insect and anti-fungal treatment. Clever insects for their security generally make their permanent shelter in such materials and for food interest, at suitable time, migrate to records. So far as the techniques of fumigation, they are effective but the system of sterilization employed by the organizations, is defective and need planning, which involve following points, for due consideration:

(i) For complete eradication of insects from an area, the anti-insect treatment, simultaneously should be given to all the records of a stack, covering and storage materials and the stack area itself.

(ii) If piecemeal system of fumigation is employed, the fumigated records should be housed in a well sterilized stack instead restoring them in their original places. The process of fumigation should be continued till whole of the records of the stack are fumigated. After all the records are given fumigation treatment, stack area including storage materials should also be given anti-fungal and anti-insect treatment.

(iii) If possible, in big Archives and Libraries, a mobile mass fumigation technique be employed so that the records holdings of their own be fumigated and if needed that may be used to fumigate the records of other organizations as well.

PREVENTION, DETECTION AND CONTROL OF TERMITES

Detection

Detection of sub-terranean termites, responsible for damaging archive-library wealth, is important step to be considered before controlling them. Periodical inspection of walls, floors, wooden frames, etc. particularly cracks, and hollow wood, are considered the direct sources to find termite infection in a building. But where such sources are not available the existence of infestation is betrayed by swarming flights. For facilitating such detections, it is necessary to leave sufficient space between racks and walls, as suggested earlier.

No two books show the same pattern of destruction but sub-terranean termite injury can be identified by the thin deposit of mud by which termites line the cavities eaten out in books.

Prevention and Control

For preventing termites infestation in a building, following measures have been found very effective.

The materials vulnerable to attack by termites should be stored at least 15 cm. above from the floor height and 30 cm. away from the wall and roof. The use of wooden racks and almirahs should be avoided for storage of documentary materials. In place of these steel racks should be used. However, if it is not possible to discard the wooden racks, immediately protection can be provided by putting their legs in vessels containing a mixture of coaltar-creosote in kerosene in the ratio 1:2. Since the mixture is inflammable, it requires precautions against flame, electric sparks etc. As an alternative to the mixture, strong solution of phenyl in water may also be used. Stone of suitable diameter accommodating legs of the racks or almirahs are suitable containers for filling mixture/solution, etc. It would be difficult to manage suitable support for legless racks and almirahs. As such, application of following varnishes on the surface of almirahs and shelves, etc. on which documents are stored, discourages termites access.

(i) 20% solution of zinc-chloride in water.

(ii) Three successive coatings of hot creosote. It is cheaper and effective anti-termite treatment but has some disadvantages. Firstly, it imparts black colour and secondly, it is a bit sticky. The creosote not be heated in direct heat but employing indirect heating technique.

(iii) A varnish prepared from components of naphtha, orthochlorophenol and petroleum in ratio 1:1:40 is a suitable coating material repelling termites. Two coating of this varnish on the shelves where records are kept is adequate to keep termites away from records.

The squeezing of above varnishes into cracks and joints and their all round application prevents termites attack. Such treatments should be repeated periodically.

For controlling termite, Poonam Chand of State Archives, Bikaner, used 40% solution of mercuric chloride in alcohol by warming over water bath. The 50 milliliter of solution was mixed thoroughly in ½ kg. of white paint and it was coated on the termite infected places. It was observed that on the fourth day of this treatment, termites appeared on other place of the same block. But after 10 days the treated places were re-attacked by termites. When infected places were again treated with the mixture, the reappearance of termite was not seen for six months including four months of rainy season. The modest experiment seems to be quite promising.

The preventive and control measures, discussed above, do not guarantee that the storage area will remain free from any attack of termites. In case, termites are very active in the building, the annihilation measures are essential to be taken up. Use of soil poisons in determitisation of building areas with success as temporary relief is considerable. The first step for such treatment is to open the cracks in walls, debris, wood and removal of disintegrated mortar, etc. Subsequently, poisonous dust is blown into the cavities and the connecting tunnels by means of a suitable mechanical device, dust gun. If such machine is not available, the poison powder is poured onto by a spoon and treated cavities are filled in with cement or concrete. *Paris Green, White Arsenic, Fullers Earth* and *Sodium Fluosilicate* are the poisons recommended for such treatments.

The anti-termite treatment of building also facilitates controlling the termite infestation. Though, the process of treatment is expensive, time consuming and requires meticulous planning, yet results are a bit enduring than the treatments referred to above. The soil poisons have been tried in USA, but, viewing over the safety measure only two chemical solution, coaltar in light fuel oil (1:2) and 5% DDT in kerosene or diesel were recommended by R.C. Gupta. Before applying these poisons, trenches, eight to twelve inches wide and a feet deep, are dug all along the foundation walls of building under anti-termite treatment. If building has basement, the depth of trench is so chosen to go into deep a feet of the basement foundation for right effect. Any crack on the floor, wall or near the treatment area should be sealed carefully with coaltar pitch or any other suitable material. Subsequently, 2 gallons of any of the poisons, per 10 feet of trench, is poured with the help of a hose or any suitable device. Subsequently soil forming six inch thick layer is poured in the trench and another two gallons of solution is poured and covered by the earth. This is enough for normal treatment of building having shallow trenches. But, if building carries with basement, deeper trenches are dug, and following the process of successive pouring of poison and soil till the entire depth is filled. The purpose of deep tenches however, is served by digging shallow trenches and drilling the number of deeper holes in them. The system is convenient and economic.

The odour of coaltar creosote indoors remains unpleasant almost for three weeks. Mixture is irritant to skin and eyes should be handled carefully.

The result of all the preventive and annihilative measures are not enduring. The anti-termite treatment can destroy the termite manace effectively in the area but after a period when the effect of poisons dies off, the entry of termites from outer areas and setting their new colonies can not be prevented. As such, after few years, the termites infestation may restart. Permanent relief, however, can be achieved by erecting termite proof building which can be had by a slight additional cost. Concrete floor firmly jointed with walls negating cracks, a moisture proof impenetrable barrier all along the external and internal walls at least one foot above the floor level are the requirements of a termite proof building. The metallic termite shield for the purpose is favoured in USA while 3"-4" thick layer of concrete for barrier, has been used with success by Indian Engineers. But, this modern plan cannot be incorporated in all the adapted buildings being used by the libraries and Archives because of cost and other factors. The principal idea behind making a building termite proof is to make it impossible for termites to make their way from the ground to the building. And that can be achieved in the adapted building more or less by replacing wooden supports, blocks, commonly plugged into walls to those of metal or treated wood. Sub-terranean termites have been eliminated from modern termite proof buildings using steel shelving, tells E.A. Buck.

Protection of Records Against War

War has never been any-thing but wasteful and destructive to both man-kind and material. The wars of past which were fought by bows and arrows, brought out only negligible loss to the documentary material. But today wars making use of advanced weapons, definitely would bring about irrecoverable loss to the wealth. The atomic weapons, the most severe and potential tools not only bring imminent loss to the archives and other documentary heritage but the radiations emerging out of explosions, may continue adversely affecting both, the living and non-living things in post-war time too.

The modern trend of every country to stock pile atomic weapons in huge number is suicidal and may convert this human paradise on earth to a fire pit. The thing noteworthy is that during active war the capital cities are mostly targeted to attack to paralyze the instrumental Government administration and activity. The property of archives is bound to be accumulated in such cities according to existing rules for establishing an archives. Obviously, the archives in modern times are in danger due to modern wars. Since Archival Authorities are responsible for preservation of information and disseminating that to users, it becomes their duty to divert attention of world authorities to think over the problem and plan strategies against the ghostic challenge arising out of modern war so that the great riches of Archives is preserved against these.

So far as preventive measures four options are possible.[+++] (i) Preserving the world against wars that is possible by complete denouncing of war. (ii) Abandonment of use of nuclear arms in wars. (iii) To reach to an agreement by world countries not to make historical monuments and archives the target for their attack during war. (iv) Nuclear umbrella for important Archives of the World Nations.

Previous three of them are ideal and in today's circumstances of world society, it is almost difficult to practice upon. The last option, advocating for facility of nuclear umbrella to archives and other sisterly organizations of repute could really be effective solution against nuclear wars.

Europe has faced two world wars, with dominating threats of nuclear weapons. Fortunately, Asia, Specially South, including Indian sub-continent was tensionfree from such threats until recent past. But, after going nuclear by neighbouring countries of Indian sub-continent and sounding threats very often have compelled India to divert attention protecting Central Capital and important cultural organisations, for preserving memory of the Nation (National Archives of India) against such pending threats. This is the high time to think in terms of preserving memory of the nation against ghostic threats. In this context, the activities which were planned by warring countries during 1st and 2nd World Wars like Italy, Germany, Britain, etc. and the experiences they gained, lessons they learnt could be of significant importance while planning strategies against the eventualities of war.

Role of Modern and Old Buildings Against Modern War

The experiences of the countries which have faced the havoc of modern wars, tells that modern structure of buildings, how stronger they are, cannot provide protection to the housed material therein. Besker K. Born, in this respect tells, "The Modern steel and concrete of archival or library building are totally in-effective against a direct hit or near miss by a high explosive bomb. Floors of the buildings collapse together with a residue of their contents. The heat generated by thermite bombs, softens steel causes girders to buckle and one floor fall successively upon the other. The structural fallacy of open elevator shafts, open deck stacks, open stairways is too obvious to need comment and too common not to induce the greatest concern for the future. The heat generated by a 15 pound bomb is enough to melt the materials like steel as wax in the surrounding area of 15 kms from the centre of the place where bomb has exploded.

The old stone arch buildings, however, does not allow collapse of flooring easily and may be thick enough, not necessarily, to be ineffective conductor of heat confining the fire to the floor or origin. Gothic structures of Muenster in Freiburg or the Muenster in Ulm or the Muenster in Aachem stood without structural damage from bomb blast. Obviously, if future repositories are to remain above ground, restudy to their structure seems a necessity.

The location of archival depositories whether in the centre or on the out-skirts of modern cities is a subject to be considered and mooted in length. Though, it was debated by world countries but no any definite conclusion was drawn. The proposal to provide distinctive protective making for structures of cultural value; Archives, Museum and monuments was never adopted. I think, it is not going to serve a lot. For instance, a bomb (15 pound) explosion can produce quantum of heat sufficient to melt material like steel located in the surrounding area of 15 kms away from the place of explosion. Means, if archival or any building housing cultural artifacts is scared of hitting directly but other areas of city locating important organizations instrumental to military operations, running

Government or aiding economic support to the country, are hit directly, the cultural institutions are bound to be severely effected even completely destroyed, if they fall in the range. The assessment is not hypothetical. World has already gained such experiences during 1st and 2nd World War.

The archives mainly located in the north-west part of the United States in the zone of Germany in the Reinland, which was an area of major military objective, were badly damaged in the Second World War, whereas the archives, located in the South-western part, suffered only slight damage. The reason was that the latter area was non-representative of large industries, main high ways and mines. Archives located in large cities like Frankfurt, Kanel, Nuremberg, Stuttgart were seriously affected comparatively to those Archives and Cultural organizations in smaller centers like Fulda, Friedberg, Eschwega, Gelnhausen.

This account, gives the clear message that archival repositories are more secure in small places/cities where heavy industries do not breath, and in the areas away from mines and highways or centers of strategic importance. The obvious reason is that the war time objective of every hostile country is to destroy the backbone of enemy nation and paralise its memory. For the purpose the centres boosting economy, aiding military operation and administering country's administration are mainly targeted. In today's perspective when two nuclear power are located in our neighbour, the great need is to think over the issue, as to where the archives (National Archives) should be located whether in the capital place of the country or somewhere else. Is there any need to make a change in principles of present archival establishment and their preservation?

The another question is to what actually be done to protect the wealth during war time. In this connection, Ester K. Borne, in his article, "Archives and War", advocates for a meticulous planning. Though many voices of despair, "How can you plan when your plans yet turned topsy-turvy" and "All human plans and projects come to naught", talk against the planning, yet notwithstanding, the rich rewards reaped of good planning are no less encouraging. For instance, the north-western areas of Germany which has survived against Second World War is the result of good planning. The planning involves the knowledge of possessions in existence and the plan for what to do with it. The classification of materials, proper packing, transport, location of suitable evaluation locales and the over-seeing of temporary repositories are worked out. Evacuation of repositories to safe places has played a very effective role in protecting wealth against wars. It is a time tested system of protecting records and other cultural materials during war. Remember that the lost cannot be recovered but let us save what remains. Not by vaults and locks which fence the records from public eye and use in consigning them to the waste of time but by such multiplication of copies as shall place them beyond the reach of accident can save the records against wars, quotes a journal of documentary reproduction 5:130 (Sept. 1942).

No doubt evacuation is highly esteemed system for protection of records against wars. But it involves various things to look into like testing suitability of sites before material arrive, from stand-point of temperature, humidity and other factors damagial to records. Opinion differs widely on the type of sites, but for long term storage, one which exists completely in isolated rural surface structure or sub-surface repositories is most favoured.

Deep shelter policy adopted by the National Gallery, London, during World War II, to save their culture materials had reaped a rich reward. However, the problem faced for was not less painful. For instance, to find a subterranean home to place the great national collections, Mines, Quarries, Caves, Tunnels and Deep Depiles capable of being artificially roofed and reinforced were visited and discussed. Site after site, was rejected on the grounds of possible flooding, insecure roofs, possible noxious gases, and difficult access, etc. At last, a site offering sufficient space measuring 200' × 300' to accommodate whole collections and possessing impressed to consider deep shelter repository. The survey of the cave revealed the facts that temperature within was 8°C and humidity near 95% to 100%. The access was not easy. However, it could be made suitable after fairly heavy work. In the circumstances, there was no alternative but to accept these figures as characteristic. It could be made ready for occupation and the National Gallery thus possessed of a repository offering a cover against aerial attack to a degree approaching impregnability.

The other problems from the stand-point of preservation in evacuation as recorded by Ester K. Borne, out of his many experiences, which focus attention for serious consideration are caused by insects, moisture, mould, rodents, salt, water, fire wilful destruction by troops, accidents and plundering. These are the sources of destruction or damage other than resulting directly from combat. There is, however, a general agreement of the need and effectiveness of some type of planning, which includes the points of establishing evacuation priorities, packing of rare items, evacuation of upper floor, selection of evacuation sites, stockpiling of boxes, co-ordination and early implementation of plans.

Duplication of Records for their Security

The international organization, UNESCO. seems not at all untouched of the problems, of security of records against war. In the post-war years, the organizations urged all countries to make photocopies of their greatest cultural treasures and to deposit them in the repositories widely dispersed through out the world. The contents of resolution item D25 of Fifth Session, Paris, 1950, reveals the fact of giving a call to member states to establish, maintain or complete a photographic documentation of their monuments, work of art and other cultural treasures to promote the exchange of this documentation and to encourage the setting up of a number of depositories, in which the reproduction of the most representatives and valuable works may be collected.

The observations of H.E. Bell, incorporated in his article, "The Protection of Archives, some lessons from the War in Italy" are very useful to those who are planning strategies for preservation of records against war. The lessons of war period are relevant to archives administration in time of peace.

The first and foremost requirement of great desirability is to have for each country a single general list of archive holdings, arranged typographically. Under war time conditions, in Italy, such a list was the *sin qua non* of protection. For the purpose it was essential that the general orders issued to military formations to respect archival repositories should be made specific in their application.

The second lesson, however, talks about the proper arrangement of record. War time Italy had experienced the advantage of it. The ancient records suffered less loss during war than the modern archives. Talking to the reasons behind, one was related to physical strength while another of psychology. The bulk of ancient records, though often great, was physically strong and manageable. Well arranged and disposed by Professional custodians, their appearance commanded respect were that much less liable to casual dispersion. Modern archives on the other hand formed usually indefinitely larger and not well arranged giving the impression to the layman soldier or civilian simply a mass of old paper, the preservation of which was no matter.

So far as India is concerned, in view of the facts discussed above, the most effective solution to preserve national memory, housed in National Archives of India, against any eventuality of nuclear war, is to establish an archives adopting deep shelter policy, in the region centrally located the country. The area of Panchmarhi, in modern Madhya Pradesh, being represented by heavy mountains of Vindhyachal series, climatically sound, in isolation but well connected with road transport, appearing in the international tourist map, possess ample possibilities to be a most favourable place for such archives where originals or duplicate whatever are deemed fit may be kept survived against the most disastering giant, nuclear war. The Rome was not built in a day, neither it could be made even, so this is the right time to think over the issue. Though National Archives of India is developing a microfilm archival repository in its Regional Centre, Bhopal. The place being located in the capital place of Madhya Pradesh and being represented by various establishment boosting nation's economy and military power hardly seems the ideal place for Archives Protection against modern war.

REFERENCES

Basu Purendu, "Why Preservation". *The Indian Archives*, Vol. No. 1, 1947, para 20.

------, "Enemies of Records", *The Indian Archives*, Vol. IV, No. 1, 1950, pp. 7-13.

Bell H.E., "Protection of Archives: Some Lessons from the War in Italy", *The Indian Archives*, Vol. IV, No. 2, 1950 pp. 164-66.

Borne Easter, K., "Archives & War", *The Indian Archives*, Vol. IV, No. 2, 1950, pp. 150-59.

Chakravorty, S. and Majumbdar P.C., "A Note on Sulph-Arsenic", *The Indian Archives*, Vol. No. 2, 1947, pp. 114-17.

------, "The Effect of Gammexane on Paper", *The Indian Archives*, Vol. II, 1948, p. 55.

Davis, John, "Preventive Conservation of Archives", *Conservation of Traditional Records Paper and Allied Materials*, pp. 71-77.

Dutta, P.K., "Conservation—The Potentialities of Neem Oil for Controlling Insect Attack", *Conservation of Cultural Property*, Newsletter No. 14, 1986, p. 7.

Guha Roy, K.K., "Library of the Indian Agricultural Research Institute". *The Indian Archives*, Vol. 1, No. 2, 1947 pp. 122-25.

Gupta, R.C., "Building & Storage Equipment Congenial to Long Term Preservation of Paper", *Conservation of Cultural Property in India*, Vol. X, 1977, pp. 29-32.

------, "Effect of Clenofix and Other Common Insecticides on the Durability of Paper", *The Indian Archives*, Vol. XVI, December 1966, pp. 29-34.

Iiams, Thomas, M., "Preservation of Rare Books and manuscripts in the Huntington Library", *The Indian Archives*, Vol. 1, No. 1, 1947, pp. 31-40.

"Indian Standard Code of Practice Relating to Primary Elements in the Design of Building for Archives" (IS: 2663-1964).

Kathpalia, Y.P., "Sterilisation of Records Against Fungus, Insects and Rodents", *Conservation of Traditional Records Paper and Allied Materials*, pp. 94-97.

Kimberly, Arthur, E., "Recent Developments in Records Preservation", *The Indian Archives*, Vol. III, Nos. 1-4, 1949.

Kishore, Ranbir, "Housing Requirements for Conservation of Archives", *Conservation of Traditional Records Paper and Allied Materials*, pp. 78-80.

Mckenny, Hughes, A.W., "Protection of Books and Records from Insects", *The Indian Archives*, Vol. VII, No. 1, January to June 1953, pp. 41-43.

Mehta, C.P., "Use of Naphthalene as Fumigant for Books and Manuscripts in Libraries and Record Depositories", *The Indian Archives*, Vol. VIII, No. 2, July-Dec. 1954, pp. 164-65.

Mitra, Rajendra, Lala, "A Report on the Indian Manuscripts in Indian Repositories", *Indian Archives*, Vol. II, 1948, p. 34.

Nair, S.M., "Biodeterioration of Paper and Other Allied Materials," *Conservation of Traditional Records Papers & Allied Materials*, p. 85.

"News Note on Preservation of Archives", *The Indian Archives*, Vol. IV, No. 2, 1950, pp. 242-43.

Prajapati, C.L., "Archivo-Library Materials their Enemies and Need of First Phase Conservation."

Rawlins, G., Ian, F., "The National Gallery in War Time", *The Indian Archives*, Vol. IV, No. I, 1950, pp. 24-34.

Ronwal, M.L. and Chatterjee, P.N., "The Control of the Indian Book Worm Becthe", Gast Rallus Indiacus Reltter, *Indian Forest Records*, Vol. 8, No. 6, 1952.

Sarkar, N.N., "A New Sterilisation Process and Standardisation of Lethal Dose for Insect Eggs. etc. by Vacuum Fumigation Chamber", Conservation of Cultural Property, *Newsletter*, No. 14, November 1986, p. 5.

Sinha, S.N., "Relevance of Conservation to Archives", Conservation of Cultural Property, *News letter*, No. 15, January 1988, p. 11.

Talwar, V.V., "Fumigation Techniques for Sterilisation of Paper", *Conservation of Cultural Property in India*, Vol. VIII, 1975, pp. 46-49.

———, "Trends in Pest Control", *The Indian Archives*, Vol. XLII, Nos. 1-2, 1993 pp. 129-31.

1+ **Soda-Acid Type**

Fire extinguishers make use of sulphuric acid and sodium carbonate. When handle of the extinguisher is striked against hard surface, acid and sodium carbonate Mix-up and reacts forming carbon dioxide and water.

2+ **Foam Type**

Fire extinguishers make use of aluminium sulphate, sodium carbonate and turkey red oil. On hydrolysis, ammonium sulphate produces sulphuric acid which in turn reacts with sodium carbonate and forms carbon dioxide and water.

++(i) Size of proceeding vols. housing in NAI (without binding) Varies from 26x39½ cm.

(ii) Size of Ocs varies from 21 x 36 cm to 29 x 46 cm.

(iii) Size of Index Vols. Varies from 20x31 to 42x33 cm.

(iv) Size of some vol. of Military Index is 27 x 50 cm.

+++ To add to preventive measures the ideal way is preserving the world against wars and that is only possible by complete denouncing of war. But it is impossible to bring into practice due to various reasons. Second alternative is, the abandonment of use of nuclear arms in wars. Again, it seems impossible to practice upon because of various factors. The third alternative is to reach to an agreement by world nations not to make monuments and archives the target to their attack during fury of air attack. It will be the third impossible, as during the ferver of air attacks, it is very difficult to identify such places and to spare them. Might has little mind, and the super might which is a blind might makes a man completely mindless. So in the age of today, shadow of danger of nuclear war has started to fall upon the archives which may get converted into complete ash if the race of nuclear war gains heavy momentum. Therefore, the time has come to meet the world archival authorities on a common platform to shape out some definite policy for the security of records against nuclear wars. Nuclear Umbrella for important Archives could be an effective solution. *Archivo Library Materials* by Dr. C.L. Prajapati.

5

RESTORATION OF DOCUMENTARY WEALTH

Restoration of documentary wealth becomes essential when they are decayed and damaged irrespective of any reason. The purpose of restoring documents is to stabilize, impart strength and rejuvenate them to maximum possible extent for the future use and also for their long term survival.

So far as the antiquity of restorative work in modern India, it is perhaps contemporary to the birth of modern Archives. The principal guiding force behind the birth of modern Archives in India was the necessity of preserving records. At the time when modern Archive was taking shape, little was known to people about modern trends on preservation, restoration and difference between the two. Whatever activities, at the time, for preservation of records were performed they were mostly the part of restorative conservation scheme.

Restorative conservation, in the saga of today, has emerged as a full-fledged art and science. It requires the basic knowledge of science and skill of a craftsmanship. Without this knowledge a restorer neither will do justice with documents nor with his profession. For a person managing restoration programme must be aware about the various steps involved for restoring documents, their sequential order, effect of skipping the order of recommended practices and nature of materials to be used for the purpose. A man with technical bent up of mind and live heart is an ideal choice for managing restorative work.

Author, himself had a long experience of conservation profession. As a teacher of the subject in School of Archival Studies, National Archives of India, he not only conducted courses in the subject and taught but also reorganised restoration programme based on latest experiences and systematized affording maximum ease. The entire programme, for the ease of understanding, can be divided into following groups and sub-groups.

1. **Pre-restoration Activities**—A document received from the repository, is not directly subjected for reinforcement but prior to that following activities are essential to be performed in the Workshop/Laboratory :
   (i) Physical Cleaning,
   (ii) Numbering or pagination,
   (iii) Opening of spine and separation of sheets,
   (iv) Examination of documents,
   (v) Sorting,
   (vi) Deacidification,
   (vii) Removal of Stains, and
   (viii) Flattening.
2. **Reinforcement**—Today various techniques are recommended for reinforcement of decayed, fragile and damaged records. But one is selected according to the need and suitability of the documents, to achieve the best at minimum cost. All the techniques being employed in India can be divided in following groups and each in sub-group:
   (i) Traditional Techniques:
      (a) Full pasting or backing,
      (b) Special full pasting,
      (c) Repair with tissue,
      (d) Repair with chiffon,
      (e) Inlaying, and
      (f) Half Margin Repair.
   (ii) Modern Methods of Reinforcement:
      (a) Solvent Lamination using tissue and acetate foil,
      (b) Solvent Lamination using chiffon and acetate foil,
      (c) Machine Lamination,

(d) Encapsulation, and

(e) Heat-casting.

3. **Guarding, filling of get-in-Paper, and Binding** : It is important to point out that every activity is taken up in the sequential order as mentioned above according to need. Deviation from this order skipping one or more activities may some time, defeat the purpose of restoration. To make it more clear, let me quote an example. If a document/manuscript without determining the nature of ink, whether or not is soluble in water, is subjected for deacidification in aqueous solution, the result of such treatment would be nothing but obliteration of written contents, if ink is soluble in water. It is undesirable. Second example of this kind is related to the repairement of sheets of manuscript without their proper deacidification. If it is done and sheets are acidic, the desired results cannot be achieved. Though sticking to sequential order in treatment of documents, in some cases is not desirable, yet in most cases it is a must, failing which it may result either in loss of information or may prove time consuming and costly. A well thought out planning is essential before executing the restorative programme for better and economical results.

PRE-RESTORATION CONSIDERATIONS

(i) Physical Cleaning

Documents gathering dust and unwanted foreign material, require their cleaning as a first and foremost activity. It is carried out manually with high order of gentleness, using cotton cloth, sponge, cotton wool, brush or any other suitable material not damaging the surface. However, the process may be accelerated by using an ordinary household vacuum cleaner with brush attachment.

Mass scale cleaning of documents can be performed using Air-Cleaning equipment. The compresure of equipment having a capacity of 300-350 liters is capable to supply air at maintained rate of pressure of 3-4 kg. per square centimeter, through a blow gun. When air is blown off along the sides and edges of the manuscripts, books, bundles, the free dust is removed off from surface and carried to a chamber fitted with renewable cellulose filters which make dust particles captive and return dust-free air to the room. The trapped dust particles, settled at bottom of the chamber, are collected and thrown off. This dust, however, may also be used for analysis of solid contaminants representing the dust-mass. The filters of equipment are cleaned off vibrations.

The air cleaning equipment (see plate 5.1) suit to the requirement of big archival organization dealing with large holdings. However, the requirement of cleaning can be fulfilled by manual means. For large scale cleaning an air cleaning unit without dust trapping filters, can serve the purpose. Such unit is comprised of compresure, a tube fitted with a suitable nozzle and valve for controlling blast of air and fumehood fitted with exhaust fan. Dust laden air is sucked by exhaust fan and expelled out of the building.

(ii) Cleaning of Stuck-up Materials

The removal of firmly stuck up dust particles, pencil and finger marks is not possible by the techniques mentioned above. However, these can be removed by soft rubbers or non-abrasive erasers such as sponge rubber, soap, art gum, kneaded and synthetic erasers made of vinyl. Mr. Kathpalia, states that kneaded erasers are perhaps, the most gentle in their action, removing dust and pencil marks as easily as a blotter soaks ink. Vinyl erasers also work easily but have a tendency to pick up their own crumbs which adversely affect the resins present in paper. Another material often recommended for cleaning of paper is bread crumbs. No doubt, it cleans but if crumbs are not thoroughly cleaned after treatment, they cause staining the paper. While selecting eraser, always prefer which is time tested and known for its usefulness.

Besides, the materials like, "Opaline" cleaning bags and soaps, wall paper cleaner ("Absorene") which are generally employed for cleaning have also been found useful for cleaning documents.

The Opaline, dry-cleaning bag is a mesh bag containing a dust eraser soap or rubber. Loose dust embedded over heavy films surface is cleaned off by rubbing the bag gently over the surface. When surface of bag becomes dirty through use, it is cleaned off by twisting, shaking for re-use.

Wall paper cleaner is available in Paint and Hardware

stores. Absorene is one of the best varieties of such cleaners. The results of test, conducted by A.L.A. (American Library Association) testing effect of the erasers/cleaners on paper, indicates that common erasers, Opaline and Absorene, do not adversely affect the paper. However, the suitability of these once again needs to be tested in tropical and sub-tropical climate.

(iii) Numbering or Pagination

Pagination of files, books, manuscripts and volumes received in the laboratory is considered an essential activity. The question is why pagination? The simplest reply of the question lies on the fact that major restorative work of Book or Vol. or any file cannot be carried out without separating sheets from the sewn text. If they are separated without appropriate pagination, it would be difficult to arrange them in their original order after reinforcement as during the process of chemical treatment and repair, the order of documents is likely to be disturbed. The Number marked over the documents would help in their arrangement maintaining original order by technical staff who is almost low qualified.

In case of pagination, the pages are to be given due consideration. As such, to serve the purpose and to save energy and time, odd number system comprising of numerals 1, 3, 5, 7, 9 ..... and so on is adopted. In the system, front page of a sheet is marked with an odd number while other side of the document is left unmarked. The last sheet of the text is, however, marked on both sides. The last number being a even number differing from other numbers of the group gives impression that it is the last number of the last sheet of the text or manuscript. However, it should be underlined and marked end as shown in Fig. 5.1 (a & b).

In case of numbering, we count the number of sheets and for the purpose natural numbers 1, 2, 3, 4, 5... and so on are used. In the system, the front page of sheet is marked with a suitable number and other side of the sheet is left blank. The number of last sheet is underlined and written with word, "End". For example 200/end. It is done to ensure that total sheets of the text of manuscripts are 200 only.

Pagination or numbering of brittled sheets in pieces, however, requires special care. The pieces of broken sheets are

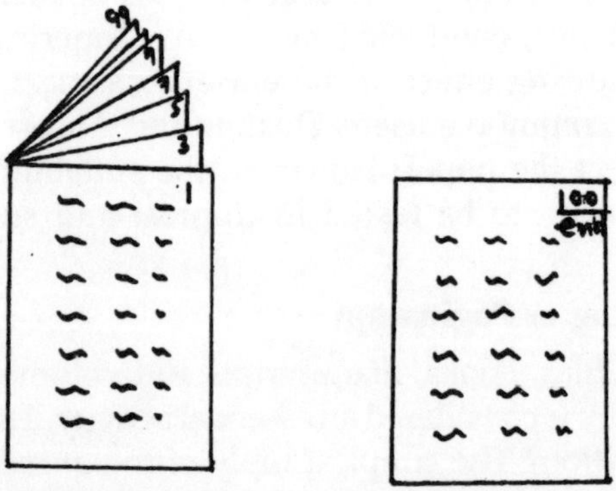

Fig. 5.1: (a) Pagination (odd no. system) (b) Pagination of last sheet

allotted sub-numbers also. For examples, if the sheet is in five pieces and the main number of the sheet is 11, the pieces of the sheet will be numbered as 11/i, 11/ii, 11/iii, 11/iv, 11/v as shown in Fig. 5.2. If more than one sheet in continuation are brittled, all the pieces of a sheet are paginated, separated and placed in a quality docket cover. Subsequently, the pagination or numbering of other damaged sheets is completed. It is essential to avoid the mixing of pieces of sheets and making the problem cumbersome.

Sometimes pagination is carried out allotting column number. It become essential when work in more than one file of similar look, nature and size is being done in workshop. Just to avoid the interchanging of sheets of files, books etc., first they are allotted column number and then paginated as per procedure. For example, if three files of same size and appearance are to be opened for chemical treatment and repair, for pagination, they may be allotted column number, A, B, C. Subsequently they are paginated, A_1, A_3, A_5, A_7 and so on. B_1, B_3, B_5 .... so on.

Fig. 5.2: Pagination of damaged sheet in pieces

(iv) Opening of Spine and Separation of Sheets

Minor repair of the damaged or torn up sheets can be carried out without separating the sheets from a text or file but major work of reinforcement and chemical treatment is not possible without separating the sheets. It requires special care as unwanted pressure applied for their separation may result in damaging them further. In case of sewn files, if threads are trimmed and the cover is removed, sheets may be separated with ease. However, in case of bound books and volumes, separation is a three fold activity. Firstly, removal of cover sheet that is done by cutting sewn threads specially in case of hollow bound texts. Secondly, the glue or adhesive from the spine of the text is softened. It is done with the help of blotting paper or cotton strips larger than the dimension of spine. The text whose adhesive is to be softened is clamped in a lying press vertically keeping spine on the top and then wet strips, referred to above, are spread over the spine. Constantly, with interval of time, water is poured over the strip ensuring that it is sufficient for absorption by adhesive but not wetting the other margins, specially written portion. Depending on the climatic conditions, lukewarm or tap water may be used for this purpose. In winter, lukewarm water will give appreciative results. After, the glue is softened, it is removed with help of sharpened knife. Now, tape and thread used for sewing the text would be visible which are trimmed and gathers are separated as third activity. From gathers, if needed, sheets are separated.

"Now-a-days, special equipments for softening of glue from the spine of books and volumes. are available. The equipments for purpose make use of water steam."

(v) Examination of Documents

Examination of documentary material is an important activity in Archives and Libraries before their chemical treatment and repair. An archival document or manuscript is examined with view to collect the facts relating to colour of sheets, nature of damage, nature of ink, artwork and position of acidity. In case of printed material, they are subjected for only two, extent of damage and acidity. However, documents of forensic purpose are also examined for establishing authenticity for which dating, type of paper and ink are made the basis.

While examining the documents, attention should be focused to the points whether the document is stained by fungal growth, soiled, broken and missing parts. Besides, colour of the document, nature of scars such as holes, tunnels or both should also be recorded. If document bears any seals, that should be taken note of alongwith mentioning the condition of seal (broken dry or cracked). Some times photo print of document before its reinforcement is essential to match the condition of it in pre-and post reinforcement.

Ink

The nature of ink whether it is soluble in water or in any solvent in which document is supposed to be treated, must be known to the restorer. It can be done by layman method without involving any chemical test. For the purpose, on the head or tail of a character, a drop of water is poured either with finger or with dropper or any suitable device. If ink is readily soluble that will bleed. If partially soluble, it will take time in bleeding which is ensured by pouring another drop of water on the same place of the character and blotting it with cotton wool or blotter. Any trace of colour confirms the solubility of ink in water. Problem may come if paper is porous and fluffy because it will absorb water quickly without showing the sign of bleeding and leaving no fraction of water for blotting. In such cases water is poured repeatedly after intervals and blotted or noted the bleeding of ink.

The solubility of ink against any solvent in which document is supposed to be treated can be tested adopting the same procedure mentioned above.

Acidity

The knowledge of nature of paper whether or not it is acidic is very important aspect of restorative conservation. If the life of document is to be stabilised and prolonged, its nature must be known as acidity is cancerous for the life of document. Without its knowledge one cannot ensure the course of treatment whether the document is to be deacidified or not. It is often known by measuring pH of the document.

For the purpose three methods are deployed. First of them make use of pH indicator paper. Second one involves the use of universal solution and the third pH meter (Non-digital and digital). The use of digital pH meter is very effective for accurate results. Unfortunately, all the methods aforementioned, are destructive and seldom applicable in case of rare documents. However, a new technique which is non-destructive, making use of pH fluid has got vast popularity today and that can be used safely.

A pH range below 6.9 is considered acidic while pH range between 6.9 to 7.1 neutral. pH above 7.1 is considered alkaline. pH ranges below 3 and above 9 are termed highly acidic and alkaline respectively.

Tamas Scholz, in his article, "New Method of Determining the surface pH of paper by Plain Glass Electrode", has mentioned that pH of the paper can be determined by making surface of paper wet without solution of it. However, the method in spite of being non-destructive can be used by experts because of complexity involved in the technique.

Dating

Dating of document is sometimes required to answer the problems relating to authenticity of a document. Though such problems are much more related to forensic work, yet they may be puzzling to conservators and Managers in Archives. Fibre composition, filling materials, colouring matters, types of inks and migration of iron in a paper normally form the basis for dating of a document. Besides, paper making technique, water mark, printing process are also used as a tool for determining age of a document.

The problems of establishing the date of origin of specimen of paper are seldom solved without controversy state Julius Grant. However, knowledge of fibre content in a paper helps in dating the documents. Carter and Pollard, the pioneers of the analytical method determining the kind of fibre present in paper, state that prior to 1800 A.D., the principal constituent fibre of paper was rag fibre (that was mainly produced from cotton or linen) and the paper was a product of cottage industry, "hand made". At about 1804, A.D., the production of machine made paper was started. A machine made paper is easily recognized. In 1840, the straw paper was produced while in 1860, the mixture of straw and rags was employed for producing paper. From 1861, after the discovery of esparto, the use of mixture, declined and the new material played major role in paper production. Similarly, another type of mixture made from esparto and rag, was prevalent for paper making during 1861-1890.

Different processes of reduction (Mechanical and Chemical) were employed for pulping of woods (hard and soft) moulding into paper. Microscopic examination of paper, though do not distinguish wood as well as process employed for reduction of them but fluorescence microscopy in Ultra-violet light could do it successfully. If the possible date when from woods of different kind reached paper industry are known, the results of fluorenscence analysis may help in dating of paper. Bamboo fibre of Indian origin was introduced for paper making after 1930.

Various filling, loading and sizing agents, besides fibres are also constituent of paper which were used from time to time by paper makers producing paper of different characteristics. Clay like materials had been added in earliest paper but use of titanium white started from 1930. There seems little doubt that rosin (a sizing material) was first used about 1800 but its use was confined to Germany until 1835.

Colouring matters which were added in pulp stuff to improve appearance of paper are also of dating value. The commonly used iron blues date early 18th century while synthetic dye stuffs late 18th century. Ultramarine was first introduced on commercial scale in 1828.

The water mark is distinctive mark incorporated in paper at the time of its moulding. Various symbols, about 100 in number of natural origin and geometrical designs have been used by different countries in different periods and thereby these make a basis for dating of paper.

Paper making techniques which have left their mark on the physical structure or character of paper, also have considerable value for dating. The use of beater instead of stamping mill for pulp preparation is an important dating characteristic. It was introduced in 1670 A.D. on the continent for which little other dating evidence is available.

The title page of a book and printing process are also the prominent sources forming inductive and deductive methods of analysis and dating a document. Title pages before 1470 were label title pages but in the subsequent period, it was adopted in colophon form. So far as printing process, the earlier books are block books printed from wood cuts, known as xylography block books. This process was introduced between 1460 to 1480. In the subsequent period, wood being unsatisfactory, was replaced by metals. Later on, by middle of 16th century A.D. metallic founts were deployed. The printing quality of wood founts and metallic founts is so different that they can be identified easily by an experienced expert.

On the basis of these facts, a history sheet of manuscript or text or file is prepared. It helps in deciding the course of action for chemical treatment and repair. This sheet also helps as a source material for research in time testing the process and effect of chemical used for treatment of document.

(vi) Sorting of Sheets

To facilitate the chemical treatment and reinforcement, entire lot of a text is divided in following three groups and each group into sub-groups. The documents of each sub-group are kept separately in quality file covers.

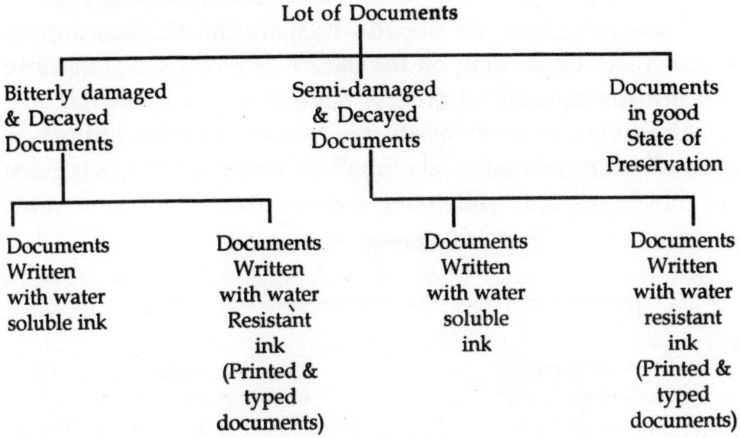

Chart 5.1 : Sorting of Documents

(vi) Deacidification

To prevent acid catalyzed hydrolysis of cellulose and to stabilise the paper, deacidification of documents is very essential. The process by which acid contents are removed or neutralized, is known as deacidification. Barrow in early forties while conducting experimental studies to find out the inner cause of paper disintegration, observed that the built in acidity of paper had greater adverse effect on document than atmospheric contaminants, and therefore, he recommended deacidification of such documents to prevent further decay because of acidity. Subsequently, in 1969, Smith took up studies to review the problem and observed that the requirement of pH value for permanent paper has been changing in different times.

The studies on requirement of pH for permanent paper, seems divided in three streams. According to first stream, the most suitable pH for permanent records is 7 while second advocates for alkaline paper having pH 9 approximately. The third stream, however, pleads the favour of slight acidity in paper say about 6.5 as it prevents the document from fungus attack. Let it be, but no one has advocated the highly acidic paper with pH range 4.5 and below suitable for permanent retention of records. As such, deacidification of acidic records, is desirable to estabilise their life, sooner the better.

For neutralization of acidic contents in paper, various processes have been developed which are chosen according to requirement depending on the nature of records. Taking into account the capacity of process neutralizing mass of records, at a time, they may be classified in two groups; deacidification in small scale and mass scale. Small scale neutralizing processes are mostly manual while other is mechanical.

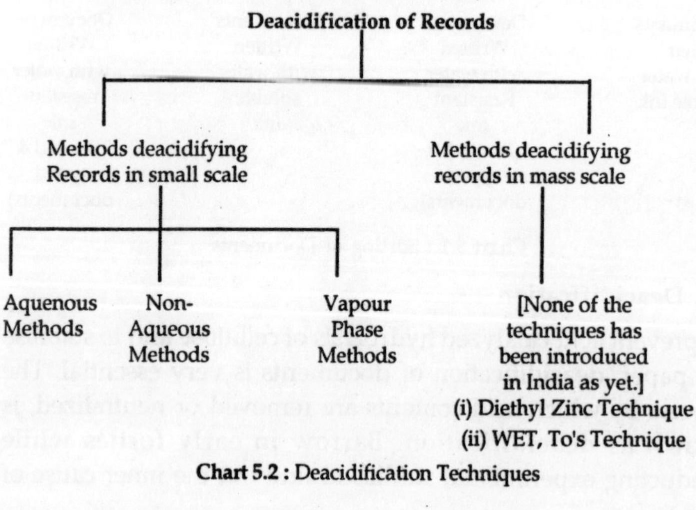

Chart 5.2 : Deacidification Techniques

Aqueous Methods

Numerous are the techniques of this category. Some methods make use of single solution while some mixture of solutions. Some are, however, such that they make use of two solutions in which documents are treated separately. These methods are useful for deacidication of records written with water resistant ink. Otto Schierholtz Method, Barrow's Techniques and Langwell Process are few well known techniques of this category widely being used for neutralization of records.

Otto Schierholtz Method

This method was invented by Otto Schierholtz of Ontario Research Foundation in Toranto, and got patented as early as 1936. The technique is based on the chemical treatment of records either by immersion or by spray in aqueous solution prepared from Barium, calcium or strontium bicarbonate or hydroxide. Because of this treatment, carbonate of respective metals deposits over the surface of treated documents.

Schierholtz, further recommended the use of more concentrated bicarbonate solutions of above metals. It is possible to prepare solutions by passing carbon dioxide at high pressure. Solutions of higher strength are useful for neutralizing acidity from ground wood paper like newsprint.

Method of Calcium and Magnesium Carbonate Solution

Mixture bicarbonate solution of calcium and magnesium is prepared mixing 1.5-2 gm. of calcium carbonate and 15-20 gm. of magnesium carbonate in one liter of water and passing carbon dioxide for about two hours. After the un-dissolved particles are settled on the bottom of container, the clear solution is decanted in a separate container and used.

The document to be neutralized from acid contents are immersed into solution for about 20 hours preferably overnight. Subsequently, excess of mixture solution is drained off and air dried. Documents to be treated in solution should be covered with waxed or oil paper or clamped in bronze, copper or plastic wire mesh before immersing in the solution. This technique will facilitate lifting the treated document even decayed, which pose problem in handling, without further damage.

On exposure in air, bicarbonates of calcium and magnesium absorbed in the paper, being unstable, are converted into insoluble carbonates which hinders documents from deleterious influence of acid contaminants, present in the paper and in the environment. Since the process is time consuming, it is not suitable for neutralizing archival documents. This process was used by Barrow nearly 1957.

METHODS OF SINGLE SOLUTION

(i) Calcium Bicarbonate

Deacidification by using calcium bicarbonate alone, as immersion technique has given promising results. However, ground wood pulp papers do not show much improvement in pH beyond 7, and suffer from yellowing immediately after treatment as studied by the author. Anne Lilnardy and Philippevan Damme in their studies on "Practical Deacidification" have not recommended this method. However, from stand point of retaining intensity of ink and colours the use of this techniques is safer.

(ii) Magnesium Bicarbonate

This process was first time used by Gear in 1957, who was working at that time in United States National Archives. At the same time, Barrow who was working on the problem of stabilization of modern book papers, tested the mixture of calcium and magnesium bicarbonates, but he preferred magnesium bicarbonate solution only. He employed the same technique used by Gear involving the immersion of individual sheets in the solution. Spray techniques has also been found satsifactory.

Magnesium bicarbonate solution of higher concentration is prepared by mixing 25 gms. of magnesium carbonate salt in one liter of water and passing pure carbon dioxide for two hours at high pressure. Magnesium carbonate is slightly soluble in water, but interacting with carbon dioxide, it converts into bicarbonate solution which is unstable. This is the reason that clear solution of magnesium bicarbonate, should be used afresh for effective action.

According to Mr. Barrow, the process limits its application relatively to strong book-sheets. The researcher, while working on the immension technique using solution, observed that neutralized papers show increase in pH that is maintained after artificial aging too. However, in some cases, the yellowing of paper and slight decrease in intensity of iron gall inks was observed. The data related to above studies is yet to be published.

(iii) Calcium Hydroxide (Lime Water)

The solution was used for neutralizing acidity of documents in large scale, as a first aid activity, at the Public Record Office, London.

Solution is prepared mixing 1.5 to 2 gm. of calcium oxide in one liter of water. It is stirred vigorously and left overnight for settling down undissolved particles of calcium oxide. Afterwards, solution is decanted and used as immersion technique. The residue of calcium oxide may further be used for preparing fresh lot of calcium hydorixde solution 3-4 times.

While studying the effect of solution over manuscripts and documents it was observed that sometimes the intensity of iron gall inks is slightly reduced and some documents suffer from yellowing in the post treatment period. However, the degree of colour dissolution and yellowing somewhat controlled by washing documents in alcohol or water alone followed by treatment.

(iv) Langwell Process

Langwell suggested neutralizing acidity of documents by using a solution prepared from disodium pyrophosphate (42 gms.), washing soda (14 gms.) and Potassium ferrocyanide (5 gms.) in 4.5 liters of water. Basically, it is inhibitor solution. Before mixing the ingredients. Potassium ferrocyanide crystals are powdered finally to facilitate their solubility as it is abit hard to dissolve. Washing soda in mixture prevents the solution from imparting greenish colour to paper in presence of iron and pinkish colour in presence of copper. It is an immersion technique.

Method of Two Solutions

Barrow's Process

This process was invented by Barrow and involves the use two solutions; calcium hydroxide, calcium bicarbonate and water.

The solutions referred to above when normally prepared have a strength of (0.15%). Basically it is immersion technique. Calcium hydroxide and calcium bicarbonate solutions are also termed as solution I and solution II respectively.

The simplest process of deacidification of documents using barrows technique is that in one of the three enamelled trys (each larger then the size of documents to be deacidified) calcium hydroxide, and in other calcium bicarbonate solution is taken. In third tray water is taken. The quantity of each solution varies according to the size of tray. Normally, two liters of solutions is adequate in a tray of size 1' x 1.5' for fully immersion of 10-15 sheets of foolscap size. Each document to be deacidified, is covered either with bifolded waxed paper sheet or clamped in batches of 10-15, between bronze/steel/ plastic wire meshes larger than the size of document as per convenience and immersed in solution No. 1. While immersing the document covered in waxed paper, take care that they are not pressed roughly as the thrust of water opposing the pressure will come into play, and because of two opposite forces, documents may suffer adversely. The simplest way of immersion is that document are first made wet by the solution No. 1 and then immersed in it by applying gentle pressure. The immersed lot is allowed to rest in the solution for about 15-20 minutes. During their bath, documents are agitated slowly with interval of time. Afterwards, documents are taken out, excess of solution is drained off and dipped in fresh water for couple of minutes to wash out excess of calcium hydroxide solution. Documents are then immersed in tray containing calcium bicarbonate solution (Sl. No. 2) and allowed to stand 20 minutes. Here also documents are agitated. Now documents are taken out, drained of excess of solution and the assembly is kept in drying rack for natural drying. Meanwhile, the fresh lot of documents is immersed in Solution No. 1 and repeating the process, all the documents are deacidified.

In one liter of solution 10-20 sheets are deacidified depending on the strength of acidity present in the document.

Calcium hydroxide neutralizes the acidic contents from document while residue calcium bicarbonate being buffer, protect the document from acidic attack within the document and from environment.

The neutralization of bitterly decayed documents in pieces, however, require special care. Only the pieces of one document should be taken at a time in folded cover of waxed or oil paper for deacidefication. Suitable arrangement for deacidefication may be seen in Plate 5.2.

The efficacy of the technique is time tested and that is why it has been accepted by various Archives world over for deacidifying the precious documentary wealth.

Preparation of Calcium Hydroxide Solution (Solution No. 1)

To prepare solution in large quantity, one container of 8 litere capacity, three containers each of approximately 30 liter capacity, calcium oxide, water and wooden stirrer (rod) are required. The containers should preferably be made of brass but made of steel or quality plastic may also be used for the purpose.

(i) First of all ½ kg of calcium oxide is powdered and transferred into a small container of 8 liter capacity.

(ii) Three liters of water is gradually added in the container containing calcium oxide. When water reacts with calcium oxide, considerable amount of heat is released.

(iii) After 10-15 minutes, more water approximately 3 litre is added with constant stirring with wooden road and solution is left for few hours to stand. When undissolved particles settle down on the bottom, clear solution is drained off and discarded. It is essential to remove impurities of sulphur etc. present in calcium oxide.

(iv) The slurry of calcium oxide is now transferred into big container of 30 litres capacity.

(v) 25 litres of water is gradually added into the container with constant stirring. Afterwards, the solution is set aside overnight.

(vi) Next morning, clear solution is decanted into another container of 30 litre capacity. This solution is calcium hydroxide solution.

The residue calcium oxide may be used for preparing fresh lots of calcium hydroxide solution 12-13 times if it is kept safe and free from any adulteration.

Preparation of Calcium Bicarbonate Solution (Sol. No. 2)

Calcium bicarbonate solution is prepared passing pure carbon dioxide in calcium hydroxide solution by constant stirring. A carbon dioxide cylinder fitted with regulator and rubber tube are required for passing carbon dioxide gas. 25-30 liters solution of calcium hydroxide require 40-50 minutes of gas treatment to get the requisite solution No. 2. When carbon dioxide is passed in solution No. 1, at first milky suspension of calcium carbonate is formed. On continuous passing carbon dioxide, a time comes when milky colour disappears leaving clear solutions. It is calcium bicarbonate solution.

(2) Non-Aqueous Methods of Deacidification

The major problem faced world over, using aqueous techniques for deacidification of records was the solubility of ink in water. Owing to this problem, it was not possible to neutralize the acid contents of a document written with water soluble ink and depicted with deligible pigments and colours. To find out a right answer to the problem researches were conducted world over. The International Institute for conservation of Historic and Artistic Work voiced a theory that a solution of alkalies in organic solvents might provide the right answer to the problem. Subsequently, in 1968, a Committee of the Institute, looking into paper problems, reported that a non-aqueous means of deacidification would not be harmful for paper, pigment and ink and may be adopted. Since then, various techniques of non-aqueous deacidification were devised and published but few below mentioned techniques were adopted widely in National Archives of India.

Bayne's Cope Method

This method got the name after its discoverer who were working in British Museum Library. His studies suggested that one per cent solution of barium hydroxide in methanol is suitable for deacidifying documents and manuscripts written with water soluble inks and also fragile records which cannot be treated by aqueous means without fear of damage. The solution is applied to the document to be deacidified either by brushing or by spraying or by immersion as per convenience.

One percent solution of barium hydroxide is prepared dissolving 1.86 grams of barium hydroxide in 100 ml. of methanol. The National museum of India experimented with the process and reported successful results using 0.5% solution rather than 1% although no test data was published by the Museum.

Its application, however, requires many precautions to be taken as methyl alcohol fumes are toxic explosive and adversely affect the colour of few inks including ball point pen inks.

Barium hydroxide, with the passage of time, reacting with carbon dioxide present in the atmosphere converts into Barium Carbonate.

Magnesium Methoxide

Five per cent solution of magnesium methoxide in methanol or in dichlorodifluoro methane (Freona12) is used for deacidification of records. Solution of this strength by weight is commercially available. It may, however, be prepared in laboratory. Solutions of magnesium methoxide in methanol upto 8% concentration by weight are stable while metastable from 8 to 11 per cent. The sheets to be deacidified are dipped into solution, agitated, removed and hung dried. This solution, however, is sprayed over documents under the treatment of deacidification. The results of neutralizing acidity are encouraging. Mr. Kathpalia, has observed that on drying the paper of treated documents cockles but to less extent than drying treatment by aqueous process. Magnesium methoxide, as residual in the documents, reacts with the water vapours present in the atmosphere forming Magnesium hydroxide which acts as a stabilizer.

Some disadvantages have also been observed. Some inks, colours and pigments are either dissolved or their intensity is adversely affected because of treatment. Some paper swells and cockles on drying. Smith has observed that instantaneous reaction of magnesium methoxide with the moisture absorbed by the paper produces a thick magnesium gel which hinders uniform treatment. These problems, can however, be resolved by controlling the conditions of treatment and duration. Mr. Kathpalia, believes that the magnesium methoxide and its reaction products do not affect the permanence of paper. However, it requires study in Indian climatic conditions based on quality and grade of papers.

Magnesium Acetate

Wilson and Forshee first time in 1959, recommended the use of magnesium acetate as an acid inhibitor in cellulose laminating films. 2 per cent solution by weight in 95 per cent methylated spirit, was found very effective to neutralize acid contents from documents. Any technique, dipping, brushing or spraying can be employed for deacidification treatment.

3. Vapour Phase Method

Vapour phase techniques of deacidification are also considered very useful for neutralizing acid contents from documents written with water soluble inks. For, the purpose, ammonia and cyclohexylamine carbonate have been suggested and used in various Archives of world.

In ammonia process, documents and books requiring deacidification are exposed in ammonia water solution (1:10) in an air tight chamber for varying time from 24 to 36 hours. The pH value of treated paper is raised to 6.8 to 7.2. Obviously, no alkaline reserve is formed over the treated paper.

Opinions are divided about deacidification effect brought out by ammonia. According to Barrow, the treated paper reverts to acidity within 24 hours of treatment and to original pH after 54 days. Y.P. Kathpalia, however, on the other hand has reported that papers deacidified as early as 1957 using this technique have not suffered adversely and the pH over a period of thirteen years has declined slightly 0.6 only i.e. from 7.1 to 6.5.

The principal reason behind the two different findings is the involvement of duration of treatment. Barrow subjected the paper for deacidification for 18 hours only while Kathpalia for 24 to 36 hours depending on the requirement. Mr. Kathpalia, has mentioned that paper of documents deacidified by the ammonia does not discolour. Paper exposed/treated with ammonia at zero temperature show better result. It is because cellulose is regenerated in ammonia and in case of textile.

It has been observed that the ammonia fume adversely affect the intensity of ink and colours of few varieties and the results of deacidification are not lasting.

Cyclo-hexylamine carbonate (dry white powder) which is soluble in water and organic solvents like methyl alcohol, once was advocated to be as good deacidifying agent as vapour phase technique. According to Langwell, 2 pounds of loose sheets of paper can be deacidified by 10 gms. of cyclo-hexylamine carbonate, contained in a box. Russian conservationists, however, stated that concentration of chemical 0.001 gms. per cubic meter is suffice to carry out, deacidification.

Inter-leaving of Books. Vols. and MSS with sheet of paper impregnated with cyclo-hexylamine carbonate solution followed by, placing them in an air tight chamber or in tightly closed plastic bags for two weeks, also deacidify them. For inter-leaving impregnated sheets are inserted in a block of 12 sheets if paper are heavily sized and coated while in case of porous and thin paper in a block of 25. Impregnated sheets should be slightly larger than the documents of books of file to be inter-leaved. Deacidification of manuscript sheets can be carried out by impregnated paper sheet placing them in closed containers.

Though findings of Barrow's Research Laboratory are indicative that treatment of document with cychlo-hexylamine carbonate prolong the life of paper, yet it brings about destruction of rosin sizing, mars brightness and causes discolouration of paper.

The treatment of paper with cychlo-hexylamine carbonate require careful handling to avoid health hazards because of its dangerous properties. At high concentrations, it releases strong alkali which may cause skin irritation and damage respiratory system as well. The odour of chemical is not agreeable. Owing to aforementioned properties, it advised that its use in air conditioned buildings should completely be avoided.

Removal of Stains

Mud, fats and oils, wax, paint, lacquer, tea, coffee, mildew, adhesive tapes and ink, etc. are common substances staining documents and other documentary materials in Archives and Libraries. The stains not only paste ugly face to the surface but also adversely affect their life with the passage of time. As such, the removal is one of the important step of restorative conservation. Before planning the removal, it would be ensured whether the document is to be deacidified or not. If so with what technique; aqueous or non-aqueous. If deacidification is to be carried out using aqueous technique, no need to attempt for removing any stains prior to as may be during the process of deacidification, the stain is cleaned off. The stains on the documents not to be deacidified by any aqueous means are, however, removed before any treatment or resizing and repair. A stain showing resistant to deacidification process, is subjected for removal employing specific chemicals, removing forcefully instead.

Some of solvents, employed for purpose, adversely affect paints, colours and inks and even sometimes leave new unexpected stain on the paper support. It is, therefore, essential to test the effect of solvents over the paper and ink prior to their application.

Few solvents found effective in dilution of stains and removing them are given in the table below:

Table 5.1: Stains and Solvents

Stains	*Solvent(s)*
1. Adhesive Tape	Carbon tetrachloride or Benzene
2. Fats	Petroleum ether or Alcohol or Pyridine
3. Glue	Warm Water
4. Grease	Toluene or Hexane or Petrol or Naphtha.
5. Ink	5% solution of Oxalic acid and 1% solution of potassium permagnate alternatively.
6. Lacquer & Varnish	Acetone or Dilute Ammonia or Methylated spirit with Toluene or Pyridine or methanol.
7. Mildew	Benzene or Alcohol
8. Mud	Water or Ammonia.
9. Oil	Toluene or carbon-tetra chloride or Hexane or benzene or Petroleum ether.
10. Paint	Mixture of alcohol and benezene or Pyridine+ or Turpentine or Amylacetate.
11. Rust	Mixture of tartaric and acetic acid of 5% oxalic acid (do not use on weak papers) or freshly prepared mixture of lemon juice and salt.
12. Sallotape	Mixture of benzene and Toluene or (sensitive tape) mixture of Hexane & Toluene.
13. Tea & Coffee	Potassium carbonate.
14. Wax	Petrol or Hexane.
15. Resins	Alcohol.
16. Shellac	Toluene, Mixture of Toluene & Benzene
17. Blood	Formaldehyde
18. Charring (heat)	20% solution of Hydrogen Peroxide.

+ : Treatment with Pyridine should be followed thoroughly, washing with water use of Benzene should be avoided.

Process

The technique, deployed for removal of stains should neither damage the surface of document nor enlarge the area of stain. Keeping this fact in view, the solution or chemical for dilution of stain is applied from its periphery to the Centre. Stain showing resistant in dilution should not be removed forcefully. The act may damage surface. In stead applying force, effort should be made to find out suitable chemical to complete the task without damaging document.

(2) Reinforcement

Rehabilitation of decayed and damaged records, is essential to correct the damages and improve their physical strength. Though, rehabilitation activities in the National Archives of India (then Imperial Record Department) had started in the period of Samuel Charles Hill (1899-1903) on the way of flattening of folded documents, their placement in new docket covers, and repair with chiffon and tissue, yet the major scientific restoration of archival material set on the post-independence period after officials of the Department were got trained in 1947 from U.S.A.

Herman Smith, in case of rehabilitation techniques of manuscripts in European Archives (Great Britain) tells that damaged paper documents are backed, patched, edged or framed with the new paper depending on the extent of the damage and the general condition of the document. He further refers three repairs; Parchment Repair, Paper Repair and Binding Repair. But in India, minor repair, sizing, tissue repair, chiffon repair (silking) inlaying, mounting, solvent lamination and machine lamination have been very popular for mending documents of different nature. Besides, some new techniques, as per necessity, answering the problems correctly, in conformity with the principles of restoration, were devised by the Indian conservations subsequently. These are half margin repair, special fulpasting and chiffon lamination.

"Stitch in time" is a very old proverb and that completely holds good in case of documentary restoration. Any mistake even of minest nature if corrected in time that prevents further damage of a book, volumes, or file or document. If such things are ignored and not attended in time, the results are contrary. Here lies the importance of Minor Repair. Documents with missing corners or mark of tear from edges or missing portion need minor repair not the major, if the paper has adequate strength. The philosophy behind doing so involves the fact that strength to a document is to be imparted at the minimum cost. The document damaged from various places, however, should be repaired fully and not partially to avoid patching and ugly look. Subjecting a minorily damaged document for full correction or repair is not judicious at all as it is not only time consuming but a costly affair also.

Before undertaking the work, it should be ensured whether the damage falls in written margin or unwritten one. If it is in the written area, the solubility of ink in water should be known and the selection of repairing material (paper and adhesive) should be made accordingly. The employment of handmade paper or alike materials would cover the written characters while use of pastes prepared in water medium, would cause bleeding of ink spoiling writing. These both, however, can successfully be employed in correcting damage in the unwritten margin. The adhesive prepared in water medium can be used in written portion provided ink is water resistant.

Skill of craftsman is of special significance as it works wonder in carrying out quality work. 5-10% solution of acetate foil in acetone or polyvinyl acetate (adhesive) and tissue paper are the materials suitable for mending a tear mark extended to the margin, written with water soluble ink. Thin maida paste or CMC Paste and tissue paper can be employed in the written margins where ink is water resistant. Handmade paper or any paper free from acidity and permanent in nature serves the purpose for fasting missing corners, filling up big holes and mending tears of unwritten margins.

For correcting tear, the document is stretched over glass top table and suitable adhesive with art brush is applied over the torn portion. Now a strip of tissue paper is laid over the tear and pressed down. The document is then turn over and repeating the process another strip of paper of equal dimension is pasted and when paste dries, the excess tissue paper is scraped off. For mending tear on the written margin, a good patch work using handmade paper or any suitable paper is equally good. Impregnated tissue papers are now-a-days available, to perform such jobs readily. Such paper may be prepared in the laboratories, using tissue paper, internally plasticized polyvinyl acetate emulsion, water and methyl or ethyl alcohol. Such papers are stored interleaved with waxed paper.

Filling of missing portion in the interior of a document invite special attention as sometimes the missing contents are rewritten by language experts and scholars with the help of knowledge or evidential sources. It requires use of filling equipment made of plywood and glass fitted on the top and an electric plug on the bottom.

Sizing or Resizing

Mechanical strength of some documents can be improved by sizing alone. But such documents before subjecting for sizing are cleaned, washed and deacidified as per necessity. It is, however, ensured that the ink of document to be sized does not run in water. Minor portions of writing such as signatures etc. on the document with soluble ink, should be protected by application of 5-10% solution of cellulose acetate in acetone or polyvinyl alcohol emulsion before subjecting the document for sizing.

Gelatine, glue, parchment, methyl cellulose and soluble nylone are some sizing materials which are employed in suitable mediums.

Three per cent clear solution of gelatine is employed hot having temperature between 43° to 45°C. It is prepared by dissolving adequate amount of gelatine in water, heating gently preventing burning and browning and subsequently filtering through fine cloth. The documents to sized are immersed in solution in enameled tray or in any suitable flat container. After sometime they are removed, excess of solution is drained off and hung dried followed by ironing. The surface grime, pencil marks etc. must be cleaned off properly, before sizing otherwise they are fixed indelibly.

Two and a half per cent solution of glue in water works as a good sizing agent. It is prepared heating glue and water to 43° to 45°C. Documents to be sized are passed through the hot size, removed after a while and drain off excess size followed by drying.

Methyl cellulose which is synthetic compound known as Glutofix (trade name) also serves as good sizer. 5-10% solution of this material in water is applied by brush over the sheet under the sizing treatment.

One and a half per cent solution of calaton (trade name of soluble nylone) by weight in alcohol works for sizing. It is prepared mixing both constituents and heating indirectly to 40°C on water bath. Documents to be sized are dipped, removed off and dried. This activity is carried out in cup board or in a well ventilated room fitted with exhaust fans to dispose off alcohol fumes properly.

Sizing of mildewed sheets, imparts adequate strength to them. Such sheets, however, first require treatment with 1.5% solution of thymol in alcohol to sterilize mildew.

Repair with Tissue

Documents not bitterly damaged, written with water resistant ink on both sides of thin or moderately thick paper are generally reinforced with tissue paper and thin maida or CMC paste.

L. Herman Smith in his article Manuscripts Repair in European Archives, says that Japanese paper (tissue) is considered less expensive and more transparent than silk gauze. It is easier to supply missing bits of writing in new ink over tissue paper than gauze.

In the first instance of repairing process, two sheets of tissue paper larger than the size of document under repair, are cut. Subsequently, the document is stretched over waxed or oiled paper on a glass top table. Now the document is uniformly wetted by means of a sponge, cotton wool or cotton cloth dipped in water. The past is applied over the document and a piece of tissue paper is so stretched over it that its portion remains protruded all around the document. While stretching tissue paper, ensure no wrincle is formed. Now it is pressed gently, with cotton cloth avoiding formation of creases. Document is now turned over on another sheet of waxed paper and repeating the process, remaining sheet of tissue paper is pasted. While applying paste on reverse side, it should be extended to the protruded tissue paper also. Repaired sheet is transferred to blotting paper and the assembly to drying rack. When dries, the protruded tissue paper is trimmed leaving a margin of 5 mm on the edge where from document is supposed to be stitched and 2 mm on remaining edges.

Repair with Chiffon (Silking)

The repair process or chiffoning is identical to the process of tissue repair. The only difference is that in this process, chiffon cloth is used in place of tissue paper. Thin CMC Paste does not respond properly in joining adherends. As such thin maida paste or thick C.M.C. paste should be employed. This technique is suitable to reinforce manuscripts and documents on think paper provided they are written or depicted by water resistant ink.

Full Pasting

This technique is known to European as backing. L. Herman Smith, has mentioned that manuscripts are fully backed if there is no writing on the version. Y.P. Kathpalia, has named it mounting but today in Indian sub-continent, it has got popularity as full pasting.

Handmade paper and maida paste are considered suitable materials for backing. But, any paper almost acid free and having permanent and durable character is suitable for the purpose. However, as a matter of fact, no any paper compatible to handmade paper is available. That is why handmade paper is preferred for full-pasting.

The documents written on one side only with water resistant ink can be reinforced using this technique.

A piece of handmade paper larger than the size of document is trimmed, moistened in water, rolled and kept aside. Document to be reinforced is stretched on the glass top table keeping unwritten side up and uniformly wetted with water using cloth/sponge. Subsequently, thin maida paste is applied over it and rolled handmade paper is so unrolled that it remains protruded on all the edges of the document without forming any creases. With wet cotton cloth, it is pressed removing wrinkles and ensuring proper adhering. Now, it is transferred on a blotting paper and assembly is kept in a drier. When dry, they are subjected for mechanical pressing in *dab* press and protruded portion of handmade paper is trimmed leaving margins as suggested earlier.

The wheat flour paste is cooked before use. In European countries, near about 1949, the paste prepared from the mixture of wheat and rice flour was employed for this repair.

Inlaying

This technique in European Archives is known as edging. Whereas it is a possible to strengthen a document from its edges, it is also possible preventing further damage because of use and handling if document is inlayed. Further, enlargement of document is also feasible using the technique. The latter one sometimes becomes essential, if documents of small size, comparatively to other documents in the lot, are found. The document of large size cannot be reduced to the lower size but latter one can be enlarged employing the technique without damaging the document and hence of judicious use in Archives, as in files, documents of different sizes are found. In case of binding, it is always desired to bring all the documents to similar size.

Y.P. Kathpalia, describing the technique suggests to cut sheet of handmade paper of required size at the first instant. Subsequently, outline of the document is traced over the sheet forming a rectangular or square geometrical figure. The interior outline reducing by 3 mm in width and length is drawn under figure and trimmed. The document is framed applying maida paste to the edges, smoothened and pressed. Actually, the process is "single sheet framing technique" useful for archival documents. But for manuscripts on thick paper sheets, double sheet framing is desirable. In this technique two sheets of handmade paper of equal size, each larger than the size of the manuscripts, are taken. Better would be if two required sheets of handmade paper are prepared by folding a single sheet. One of the two sheets is marked by the same number, over the document to be enlayed. The document is placed on this and outline of that is marked and reduced approximately by 3 mm all along the sides and interior of that is removed. See Fig. 5.3. Subsequently, a smaller window than the first one is cut on the

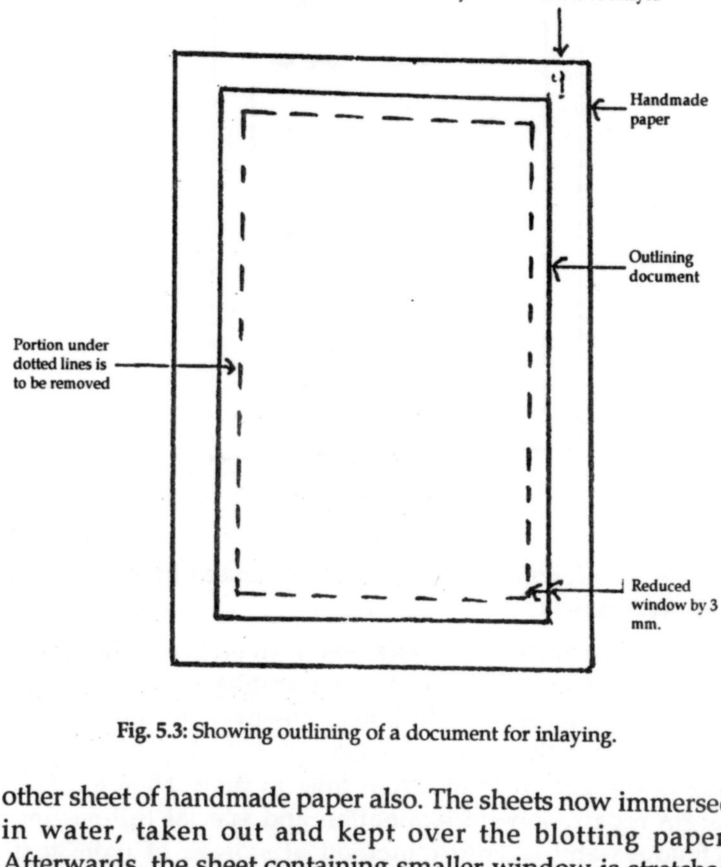

Fig. 5.3: Showing outlining of a document for inlaying.

other sheet of handmade paper also. The sheets now immersed in water, taken out and kept over the blotting paper. Afterwards, the sheet containing smaller window is stretched over the glass top table and thin maida paste is applied. Documents to be inlayed in now so fixed on the smaller window that the written contents of the document are clearly visible. The sheet of paper bearing larger window is turned over the document and pressed gently. While doing so, it should be ensured that the written margin is not blocked by the paper sheets and no wrinkle is formed. So inlayed document is dried, pressed and finished from the eges if required. The size of both the windows is reduced to give support to the document to be inlayed. See Fig. 5.4.

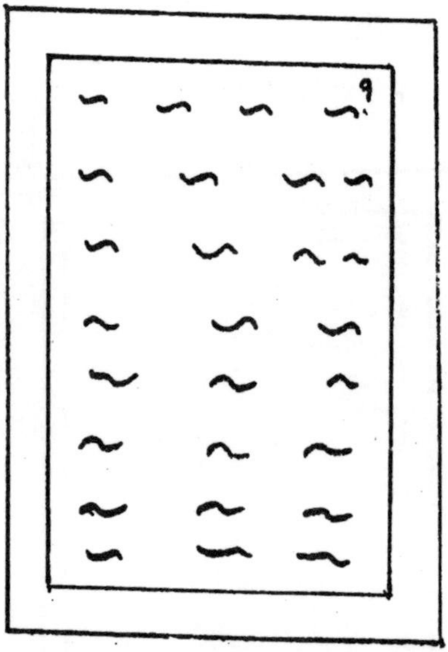

Fig. 5.4: An inlayed Document

Manual Reinforcement Techniques Invented by the Indian Conservationists

The most of manual techniques described earlier, more or less are based on the techniques being employed in European Archives. Some new techniques, however, to answer the problems correctly as per principles of archives conservation were developed in India, from time to time. They are half margin repair, solvent lamination and special full-pasting techniques and scientific method of storage of palm leaf manuscript.

Half Margin Repair

This technique was employed in India to impart strength to documents of specific character. Such documents were unusually usual, because of having almost 50% margin unwritten on the both sides. Though, such documents could be reinforced by techniques of tissueing, chiffoning and lamination, Yet these were discarded only as being costlier and failing to strengthen such documents at the minimum cost as desired under the principles of repair.

National Archives of India, State Archives and other Custodial Agencies are employing the technique where the records of specific character are in custody. In a way, this technique may be termed as full-pasting in piece meal.

Two pieces of handmade paper are cut, wider than the unwritten margin and longer than the document. Each picee is marked with the number on the sides of the document. Now one of the handmade paper sheets is stretched over the unwritten margin of the document and a zigzag figure is drawn such that pointed portion falls between the lines. See Fig. 5.5. Now sheet is removed and shadowed portion is trimmed off as shown in Fig. 5.6. Repeating the process, another sheet of handmade paper is also prepared on the basis of the reverse of the document. Both the sheets are now immersed in water, removed and kept aside. Subsequently, they are pasted employing thin maida paste on respective sides inserting pointed portions between the lines. See Fig. 5.7. Document is dried, pressed and trimmed accordingly.

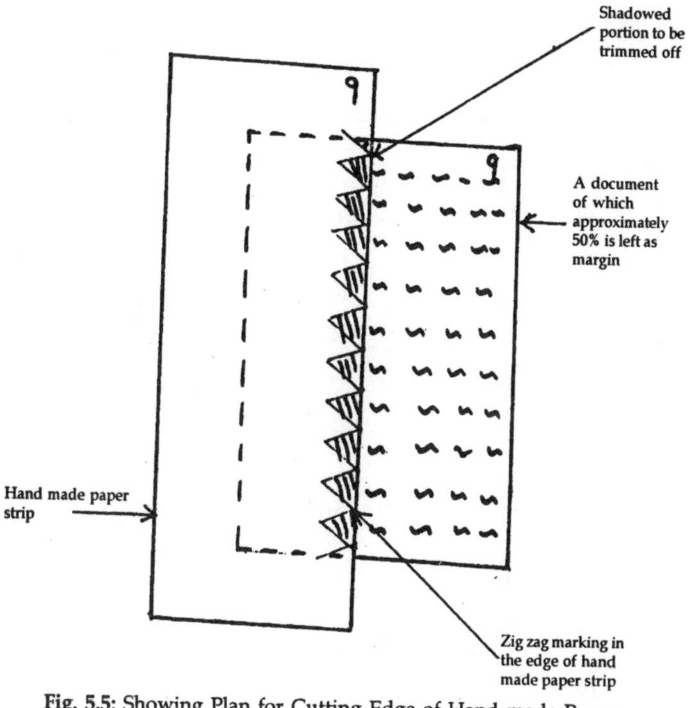

Fig. 5.5: Showing Plan for Cutting Edge of Hand made Paper

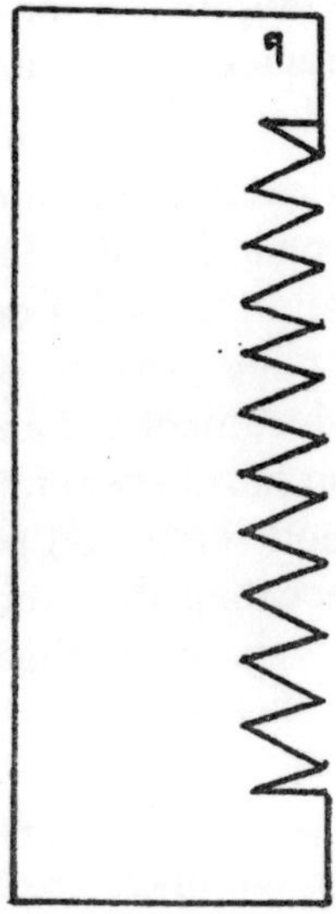

Fig. 5.6: Hand made paper strip after shadowed portion is removed

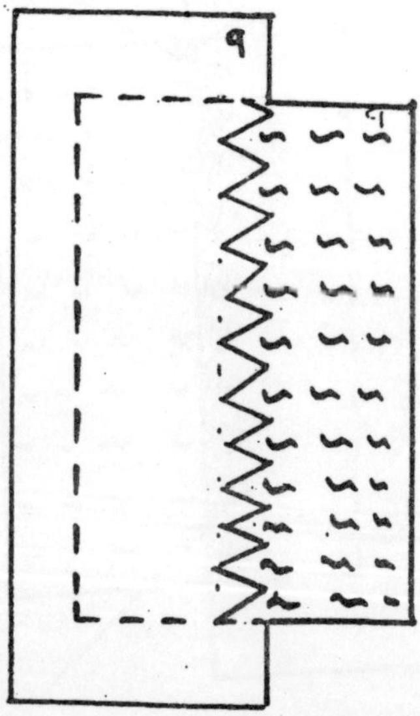

Fig. 5.7: Document half margined from one side. Similar reinforcement is required from other side also.

Solvent Lamination

Machine lamination invented in U.S.A. is extremely an old practice. It requires deployment of a machine and heating system. When the process was invented, it brought out a revolutionary changes in the field of documentary restoration in European world. The relevant factors about the technique, its merits and demerits would be discussed separately. However, the discovery of manual process of lamination called solvent lamination or hand lamination gave a new orientation to the art of rehabilitation of documentary wealth. The method, invented by O.P. Goel at National Archives of India in 1952, is a slow process but regarded as safe method for reinforcing even bitterly fragile documents in pieces which is hardly possible in machine lamination. Safe in the sense that it does not employ heat as required in machine lamination.

The document to be solvent laminated is first cleaned of foreign material, deacidified if needed and flattened. Two sheets of tissue paper and two sheets of acetate foil each larger than the size of document are cut. Out of two, a piece of tissue paper is spread on the glass top table. Subsequently, a piece of acetate foil almost equal to the size of tissue paper is stretched over it. Now document to be laminated is stretched. Acetone, soaked into a piece of cotton wool, is applied on the corners of document and centre. It is essential to fix the three togather. Subsequently, another piece of acetate foil is stretched and then piece of tissue paper (see Fig. 5.8). Now acetone (pure and colourless) is applied on the tissue paper by a cotton wool. Initially, it should be applied from centre to the periphery side tending a moderate force to expel out air captive between the document and acetate foil. The surface is moderately pressed with hand Palm or cotton. The foil, below tissue paper dissolves in acetone and serves as an adhesive joining tissue paper and document with each other. After a minute or so when the surface becomes dry, the document is taken off the glass and turned over. Repeating the process, another piece of tissue paper is pasted on this side applying acetone. The protruded tissue paper is, now trimmed, leaving a width of atleast 5 mm from the edge where from document is supposed to be stitched or bound and 2 mm from remaining three edges. As per requirement, the repaired documents are pressed in a *dab*-press for 1-2 hours or flattened by a moderately hot electric press after placing it beneath a card-board or blotting paper.

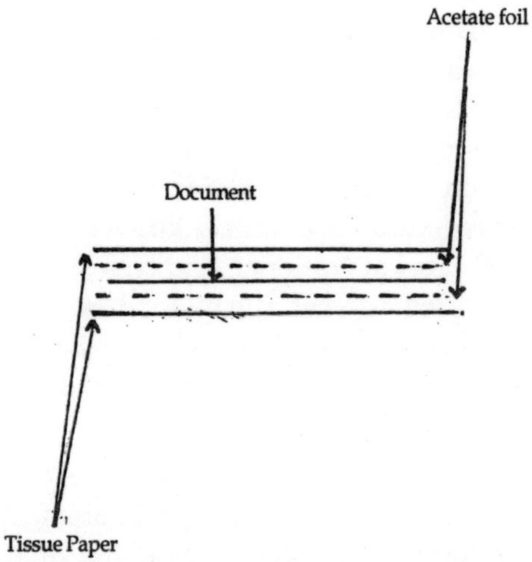

Fig. 5.8. Thematic diagram showing arrangment for solvent or any lamination.

The bitterly damaged documents into pieces, however, require special attention. The pieces are first, arranged in position on a glass surface, placing acetate foil beneath the joints. Subsequently, acetone is applied with a piece of cotton wool over the joints, ensuring appropriate fixing. Too much and too little of acetone may result in inadequate adhering. The integrated document, is then covered with acetate foil (larger than the size of document) followed by tissue paper and processed as stated earlier.

Special Full-Pasting

The technique was devised by Author for reinforcement of Tangyur Manuscripts while working in a Project, "Conservation of Tawang Monastery", Tawang, Arunachal Pradesh.

The sheets of Tangyurs are known for their unique structure. They are made of multisheets of thin papers sticking together one over the other employing local adhesive See Fig. 5.9 (a). The thin paper sheets of Tangyurs from some places, had come apart because of loss of binding strength of adhesive, with the passage of time. The sheets were comparatively thick and written on both sides with black ink, resistant to water. Because of the facts, that all the traditional manual methods of repair were not capable to give right answer to reinforce the sheets, a new reinforcement technique was devised to strengthen the MSS sheets, known as special full pasting.

The written sheets, of both the sides of a Tangyur Manuscript were first taken off adopting suitable measures and subsequently they were pasted on both sides of a good quality handmade paper in such a fashion maintaining look of original sheet [see Fig. 5.9(b)].

The system whereas fused the new lease of life to the Tangyur sheet, the work was performed on the low cost.

A document made of multi-layered sheets, can be repaired using the technique.

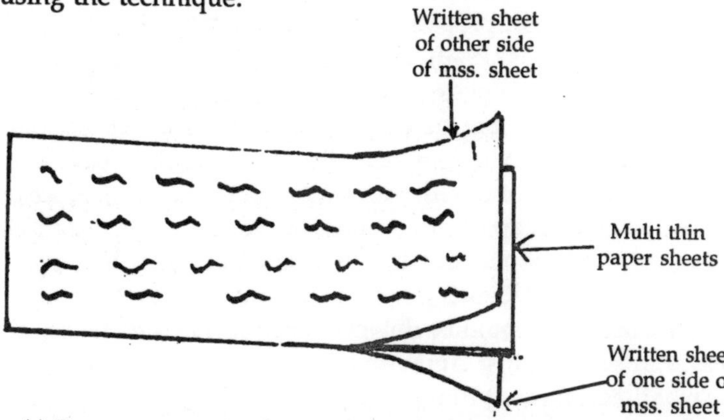

(a) Tangyur manuscript sheet made of multilayeres of thin paper sheets

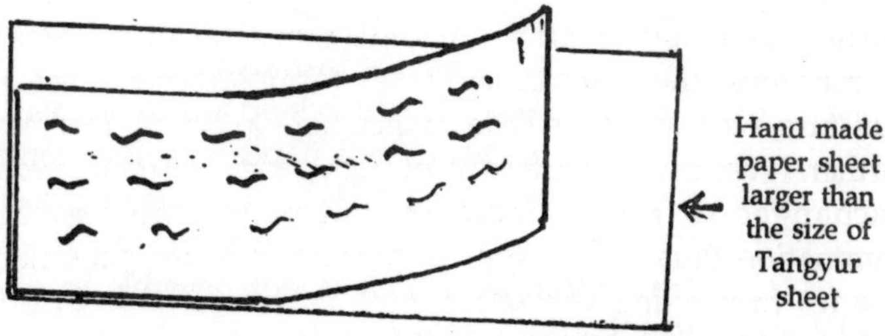

(b) Plan of pasting of removed mss. sheet on hand made paper. Another mss. sheet is pasted on the otherside of hand made paper sheet

Fig. 5.9 : Showing some steps of special fullpasting of multilayered manuscripts.

MECHANICAL REINFORCEMENT

Machine Lamination

Hot sealing of a document, deacidified or as it is, with cellulose acetate film of specific thickness, deploying suitable machine is known as machine lamination. Previously, machine lamination of documents carried out with acetate film only. But, later on when observed that it does not impart appreciable strength to document, the reinforcement was carried out using both tissue paper and acetate film.

Steam heated flat bed hydraulic press and electrically heated roller press (rotary laminator) are the early machines employed for lamination. (See plates 5.3 and 5.4). Impregater Laminator, is the device of new generation. It is being used in National Archives of India. [See Fig. 5.5].

The lamination process was developed by the National Bureau of Standards (1940), U.S.A. under the leadership of Mr. A.E. Kimberly. This process inceptionally was subjected to more criticism than any other method of rehabilitating documents. The criticism, mostly questioned the permanence of cellulose acetate (was then not tested with time), reversibility of process and the process incapable to remove the cause of deterioration inherent in the paper and ink. These objections, however, removed as the test results went into the favour of high stability and the document could be delaminated harmlessly. In case of third argument against lamination (that it does not remove the cause of deterioration) it was suggested that highly acid paper should be deacidified suitably before lamination or repairing by any other method. If a paper slightly acidic is sealed hermetically by lamination between two impermeable membranes, the external agencies (atmospheric gases and moisture, etc.) will have little chance to help acidity increasing further. Presumably, the acidity of such paper will either remain unchanged or rate of increase will be inhibited greatly. Lamination thus gives a type of protection to the documents, besides improving strength which is not possible in other technique of restoration.

Lamination of alkaline treated papers increased the life of a document further, says S. Chakravorty in his article, "A Review of the Lamination Process." Mr. Barrow's views in this respect are, "On accelerated aging, it was observed that silked paper had lost 52% of their folding endurance, those laminated but not treated for acid neutralisation had lost 31% while those laminated after acid neutralized had lost 5%. Obviously, lamination of an acid free paper, retards the rate of decay besides imparting strength.

Todays records on cheap paper does not last long as the old paper used to be of excellent quality. The rate at which modern papers decay, would not allow us to consign them patiently to their fate as we have been used to. To combat the vast accumulation of records on inferior paper that is piling on Archives and to take care of the mending work long over due, must have some mechanical means of rehabilitating the documents inexpensively and expeditiously without any skilled mender. The machine lamination is the right answer to fulfil this demand.

Before subjecting documents for lamination, their envelops are prepared employing acetate foil tissue paper and acetone. Tafflon cloth made of glass fibre is used during lamination, for covering envelops.

Encapsulation: It is the modern manual technique of reinforcing fragile and decayed documents. It is the only technique by which a document is imparted strength without pasting any additional material to the surface of document.

Quality transparent polyester sheet made of probably mylar, adhesive tape containing adhering substance on both the sides, a ruler and rubber roller are the basic requirements for encapsulating a document.

Two sheets of mylar, larger than the size of the document to be encapsulated, are cut. One of them is stretched over glass top table. The document under repair is stretched over it, outlined, and removed. Subsequently leaving a margin of approximately two millimeters from outlines, adhesive tape referred to above is pasted over the Polyester sheet as shown in figure 5.10. The document is placed within the frame of adhesive tape over the polyester sheet and another sheet of polyester is stretched ensuring alignment with other polyester sheet. The rubber roller is now rolled over the adhesive tape alaround. Thus the document is made captive between two

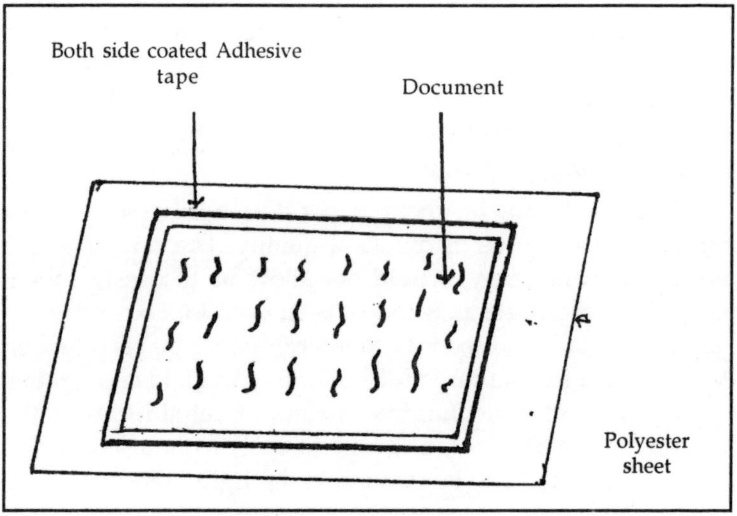

Fig. 5.10: Encaptualation Plan

polyester sheets. The electrostatic charge developed between the polyester sheets, catch holds of the document giving no chance for its movement. The encapsulated documents, paintings and other records making use of thick inks and colours, generally suffer adversely with passage of time as hair cracks develop on hue layers (ink and colours) due to electrostatic charge.

A document broken in many pieces can not be reinforced deploying the technique. An encapsulated document is protected, against light, dust and other ravages caused by environment. However, before, encapsulating the document, it should be deacidefied first.

Leafcasting: It is modern mechanical technique, introduced first time in Jewish National and University Library in Jerusalem. This technique is useful to fill up holes and missing portions in paper documents uniformly and mend them improving their physical strength. The machine deployed for the purpose is known as leafcaster which in principal composed of following three segments.

(i) Flatbed framework – Document to be leafcasted is stretched over it.

(ii) A container – It contains mixture of pulp and an adhering substance. It can move vertically and horizontally over the flat bed frame work.

(iii) Suction arrangement—It gets place in the machine beneath the flat bed frame work. See thematic figure of leaf-caster at 5.11.

The document to be leafcasted is stretched over the flat bed containing water. Calculated amount of mixture of pulp and adhesive (CMC, PVA) is sprinkled in water of flat bed containing document. The suitable amount of pulp to be sprayed is determined keeping in view the size of document and the thickness of layer of cellulose fibres to be formed over the surface of document. When suction machine is made operative, it pulls the fibre to settle over the surface of document, filling holes and missing portions, forming a layer of requisite thickness over the surface of document. This is the fundamental principle on which machine works. Today, very advance stage computerized leaf casters are available in

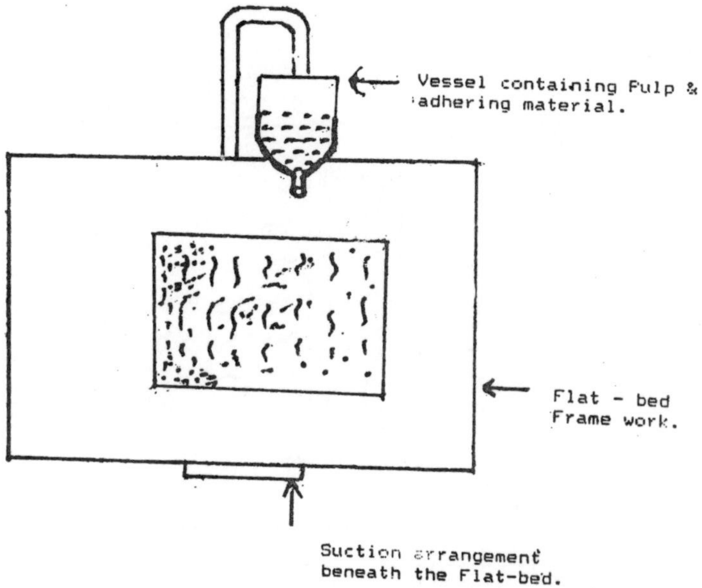

Fig. 5.11: Thematic Arrangement of Leaf Casting Machine

European Countries. But working principle of these machines is the same. Using leafcasters, variety of documents can be reinforced but they should be written with indeligible inks. This technique is very suitable to reinforce hole bearing printed documents.

(3) Guarding and Filling of Get-in-Paper and Binding

The pasting of document (repaired or un-repaired) on edges of a guard flap as shown in Fig. 5.12 is known as guarding.

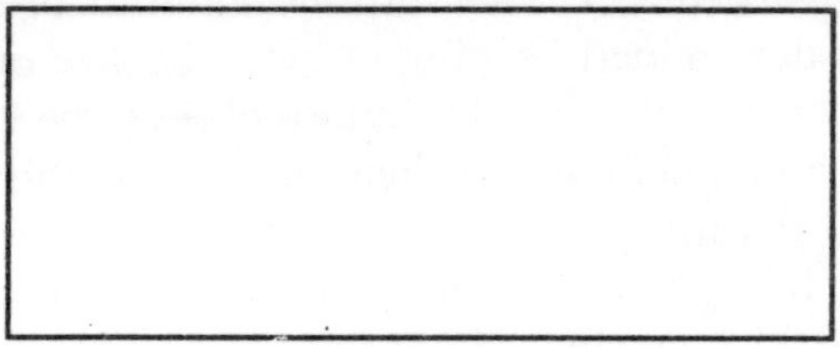

(a) Guard flap.

"It should be permanent & acid free paper. The width of guard flap is considered according to the size of document and thicknes of a Book or Vol."

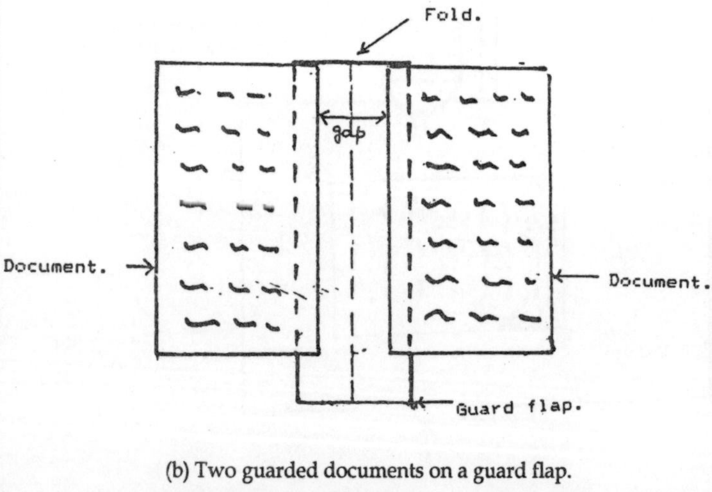

(b) Two guarded documents on a guard flap.

Fig. 5.12: Showing guard flap & two guarded sheets.

Note: Plans are available to guard eight sheets or more in chronological order to make a section or judge.

The objective of guarding is many fold. Few of them are, protection of document against stitching, freedom from stress and strain during the time of opening documents and use and matter of convenience for taking document out of the sewn text without dismantling the binding as and when required for any purpose of Xeroxing and repair etc. However, the main objective of guarding is protection of documents.

Gathers each made of eight sheets, are prepared using suitable technique and filled in by get-in paper before stitching and binding.

Binding

After chemical treatment of archival document and book sheets, their repair and guarding they are arranged chronologically, collated, filled in get in paper and stitched. Subsequently they are subjected for binding to bring them into book or volume from. But before escalating talk on other components of binding and technique, it is essential to know what binding is? It is process of collecting sheets of MSS, book or documents of a file together in chronological order and putting them between two hard covers. But code of conservation ethics does not permit to do anything on documents that hampers their life, slightly or largely. As such, sewing through documents, is not at all safer for the life of document as during the process of stitching not only their physical strength is adversely affected but also it remains at stake through out life because of the less margin on the document sewn. Normally, in a rebound book or vol. pressure over the sheets is applied for reading written contents that generates a stress and strain over the sewn portion damaging documents in the long run. Guarding is the right answer to sort out both the problems. Firstly, because of guard, stitching would be through guard paper. Secondly, margin because of guard flap on each document is enlarged facilitating opening of text without applying any external pressure. Other advantage of guarded sheets are that they can be removed from the text, for the purpose of repair, chemical treatment or Xeroxing or any other purpose without dismantling the binding and repasted afterwards to their original places. Really, it is an excellent archival technique to protect the documents economically. Unfortunately, the name of person who invented the technique is not known but he deserves every praise for his technical bent of mind and devotion to archival profession. It is a precious gift to archival world to conserve documentary wealth. While guarding the sheets, gathers, each of 8 sheets, are prepared. To match the thickness of gathers from the spine to the joint margins of document and guard flap (here thickness is almost twice to the spine), strip of quality paper known as get-in-paper, are inserted. A get-in-paper strip should be made of acid free paper, having width equal to the gap between pasted sheets on the guard flap, is considered suitable. See Fig. 5.12 b. Without filling of get-in-paper, if gathers are sewn and binding boards are fixed, they will remain open spring from fore-edge causing inconvenience in storage and will get damaged very soon.

Gather's filled with get in paper are stitched and subjected for binding. Binding of books and volumes after reinforcement and guarding is termed as archival binding. It is also termed as reinforced binding.

To understand the trade of binding properly, one should be well aware about the various parts of binding that can be seen in make up of a bound book. See Fig. 5.13 (a&b)

Types of Binding

The binding, taking into account the spine of the books and use of covering material over them is divided into two; fast back and hollow back binding. In previous type of binding the covering material (cloth or ragzine or leather) is pasted over the spine of the book or volume while on the latter it remain free. When book or volumes are opened, in fast back binding, covering material of spine moves with spine facing stress and stain. The material of spine, however, on hollow back, arches during the opening of text and remains free from strain see Fig. 5.14. Obviously, the covering material of hollow binding is more safe than fast back bound book and volumes.

Each of these bindings, is however, sub-divided into three categories; quarter, Half and Full binding. This classification is based on the use of material. If material (leather, cloth and ragzine) is used for covering the spine only, it is called quarter cloth or quarter lather binding. If the material is used on the spine and the four corners of the board, the binding gets the

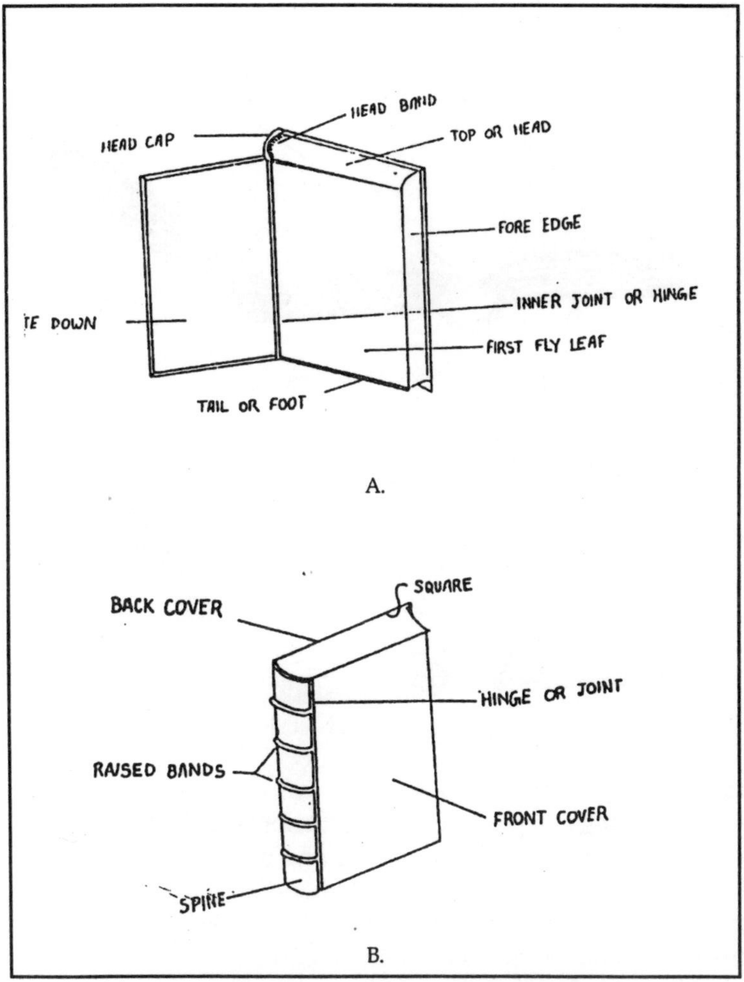

Fig. 5.13 (A&B) Parts for Hand Bound Book or Volume

name half cloth or half leather binding. Its nomenclature changes to full leather or full cloth, if leather or cloth respectively are employed for covering hole the spine, corners, front and back of boards. Besides, there are modern binding techniques such as comb binding, spiral binding and thermal binding but they are not suitable at all for archivo-library binding.

Binding of library material and library binding are two different aspects. Former means the types of binding employed for binding of library material by publishes. Some of them are

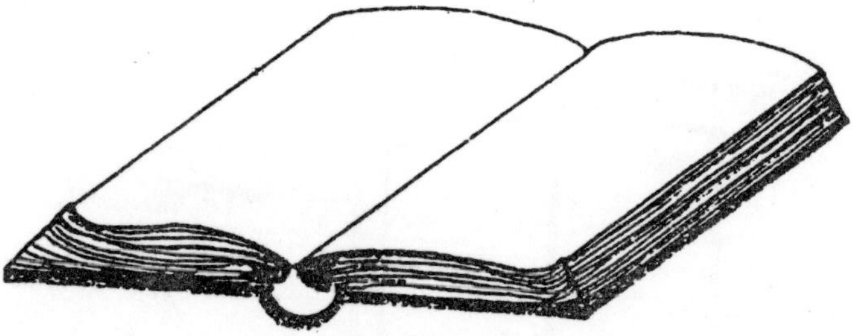

Hollow back binding

Fig. 5.14: Fast back binding

case binding, thermal binding and soft cover binding. These all the bindings are temporary and do not last long. The library binding, however, meant a durable and permanent binding. It is hard cover binding, in which gathers are sewn using cards and tapes which are laced in cover boards keeping text and cover attached to each other permanently.

After stitching of sheets of Mss. and rare books, they are attached with end sheets, followed by gluing and rounding of spine. Subsequently, hard boards are fixed and covered with binding materials of quality.

Fixing of hard board on both sides, the front and back of a book or volume and covering those with materials like cloth, paper, leather, etc. is not the process called binding. It requires more to be attended prior to fixing boards and covering them for protection of text and durability of binding. The simplest and most obviously needed protection is to glue over the spine to protect the exposed cords and sewing.

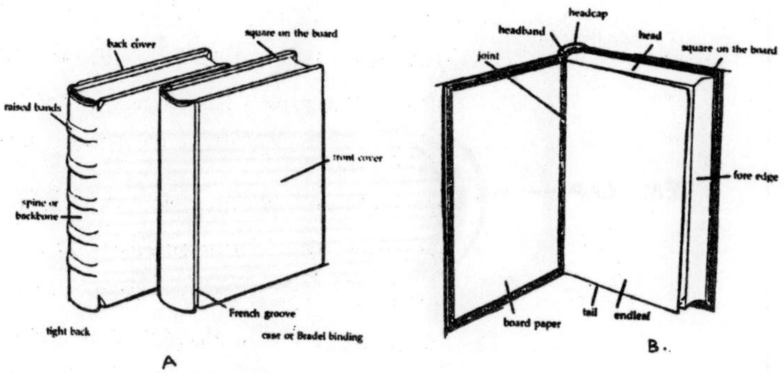

Fig. 5.15: Showing Cord laced into Board

Materials for Restoration and Binding

All the materials required for carrying out restoration of documentary wealth may roughly be classified as under:

Materials for Restoration & Binding

- Stationery Materials
  - *Materials For Repair*
    - (i) Tissue Paper
    - (ii) Handmade paper
    - (iii) Ledger Paper
    - (iv) Acetate foil
    - (v) Chiffon Cloth
  - *Materials for Binding & Mounting*
    - (i) Binding Cloth
    - (ii) Binding Leather
    - (iii) Board-Straw board, Grey board, Mill board & Split board.
    - (iv) Sewing thread
    - (v) Tape
    - (vi) Cotton cloth-coarse and fine
    - (vii) Sand paper
    - (viii) Immery paper
    - (ix) Decorative materials-Gold tooling, silver tooling, Printing
- Adhering Materials
  - (i) Sodium compound of Carboxymethyl Cellulose (CMC)
  - (ii) Starch (Maida)
  - (iii) Glue

Chart 5.3: Materials for Restoration & Binding

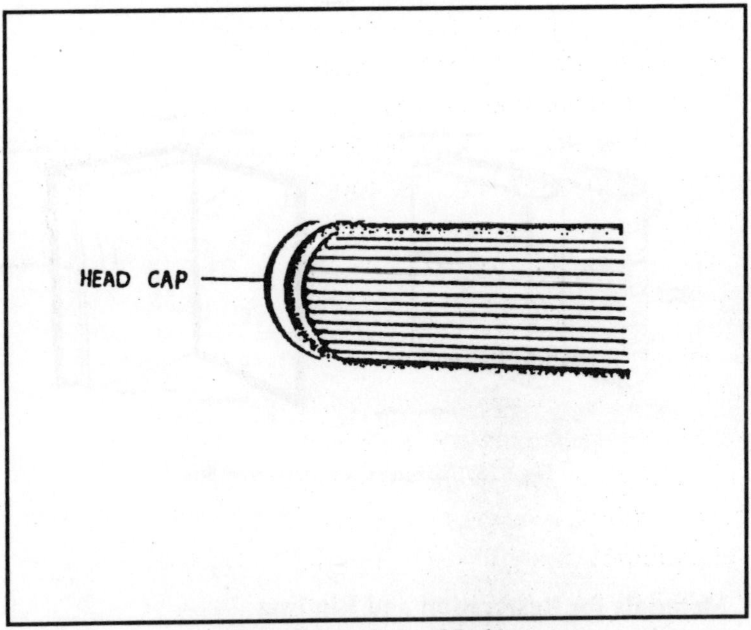

Fig. 5.16: Rounding and Backing

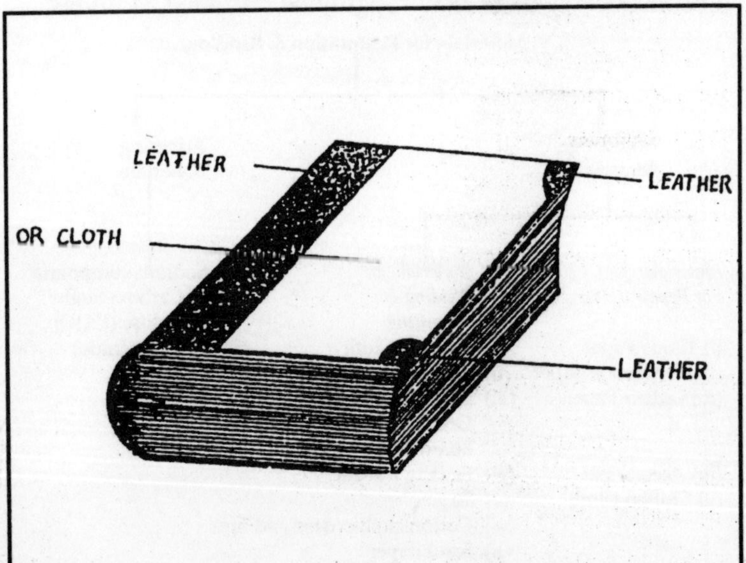

Fig. 5.17: Covering of Card board

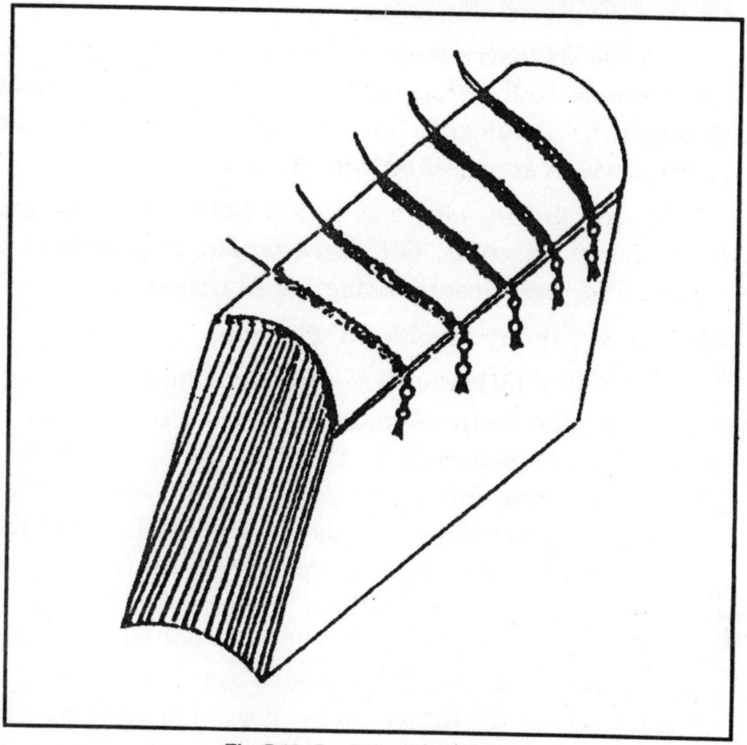

Fig. 5.18: Cord Laced into boards

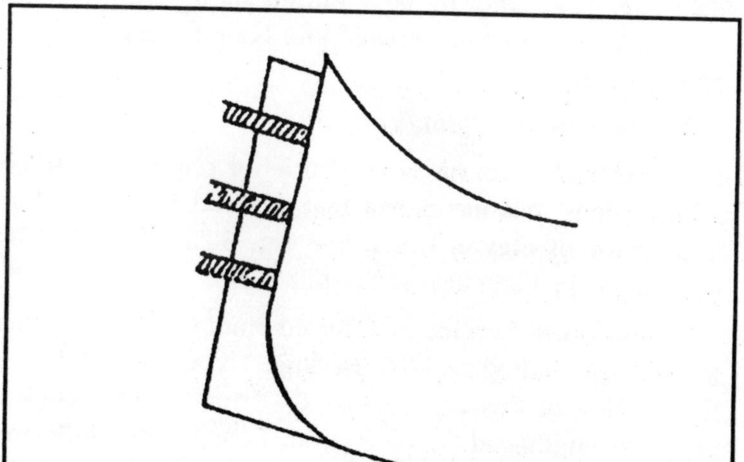

Fig. 5.19: Tapes inserted between Split Board Files

All the stationery materials, for durable and permanent repair should be free from acidity and other foreign contents damagial to documents. They should also not stain the records, which are repaired with them.

Sewing thread, tape and cotton cloth, should be free from sizing materials. For characteristic requirement of materials for permanent binding see Annexure 4.

Binding of Library Wealth—A Planning

Today we live in knowledge or information society and libraries are the source of information. Since conservation of archivo-library materials is a costly affair, a meticulous planning is a must for conservation of library wealth. While planning the programme, one should keep in the mind that all the library books are to be categorised in two groups; ordinary and rare publications. The preventive care is provided to all type of library collections but for restoration only the rare books which are treated at par to the archival documents are to be treated on the lines of archival records. However to keep the ordinary books in order for day to day use, they should be mended using minor repair techniques is a matter urgency and bound work should be corrected timely.

Chain of a Bound Book/Vol.

Dr. Ranganathan, in his book, "Social or physical Biography for librarians" has mentioned that a bound book or volume is a chain of eleven links. See fig. 5.20. IS: 5030-1965 (Reafirmed in 1969) also refers this chain.

Reinforced Binding: Reinforced Binding is very identical to archival binding or MSS binding. The unique feature of the binding of this type is that the structure of a book or vol. is strengthened properly from hinges with cloth, torn out and weak pages are repaired, maps and charts if any, are mended properly and affixed at appropriate places in the book/vol. before subjecting the same for binding.

The council of ALA, the committee of ALA and the Library Binding Institute (LBI) [The Book Binding Committee of ALA] had approved some below mentioned standards for reinforced binding in 1939.

(i) All the books for reinforced binding, shall be completely reserved in accordance with the minimum specifications for A class library binding.

(ii) All the materials used for binding shall be in conformity with the maximum specifications

(iii) A Class cover, publisher's light weight buckram and publishers original covers shall be employed for reinforced binding.

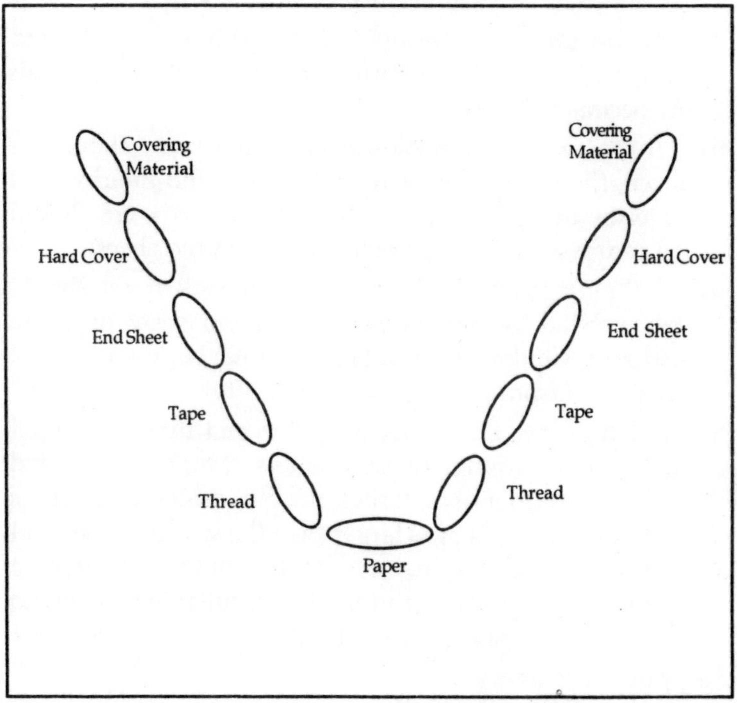

Fig. 5.20 : Chain of Eleven links

(a) *A Class Cover*: Cover similar in appearance to the publishers original cover made of heavy weight library buckram or pyrosoline coated fibres as required by the minimum specification, is known as a class cover.

(b) *Publisher's Light Weight Buckram Cover*: Similar to class A cover but made of a light weight buckram with the publisher's original design imprinted there upon, falls in the category.

(c) *Publisher's Original Cover*: A reinforced book of this type would have the publisher's original cover taken off by the binder.

Features of Reinforced Binding

(i) End sheets of all the books and vols should be guarded at the back by pasting linen or muslin strips.

(ii) All the torn and decayed sheets should be mended properly using tissue paper, chiffon or other materials of permanent nature.

(iii) All the sections torn or broken at the spine should be strengthened by fixing linen/cloth in the interior and exterior along the fold. Wrinkles, folds, warps etc. should be removed strictly without damaging the sheet.

(iv) All the Art plates, forming the part of book or vol. should be guarded by inserting a transparent sheet of tissue paper or similar material (all acid free) between the text sheet and front side of each plate.

(v) Folded plates, charts, maps, plans and other extended sheets, not forming normal sections, should be mounted deploying either of materials. muslin, chiffon cloth, linen or tough paper. Leaves larger than the size of vol. should not be pasted. All outsized sheets should be reinforced at the folds and placed in the book, suitably causing no harm to other sheets of the book.

Equipping A Bindary

Equipping of a bindary, really needs thoughtful planning failing which may up set budgetary proposals and sometimes result in wastage of money. The real requirement is the guiding force behind equipping a bindary workshop as it may vary from shop to shop. However, some tools and equipment for binding purposes are common to every bindary. The beginner, should, however, visit a well established handbook binding shop and discuss possible equipment purchases with an expert binder before setting up the workshop. Then he should decide what type of binding equipments are most necessary to accomplish the binding works in his shop or Library.

Space and money are two important considerations for setting up a bindary. So far as space, at least work space about 8' long and 4' wide is desirable for one person shop. Purchase plan should be chalked out keeping in view the budgetary position. Under tight money circumstances every purchase be a wise and sound investment.

The equipments and tools required for a well established, modern bindary for archives and leading library are many and they can be classified as under.

(i) **Leather Pairing Machine :** This machine is primarily used for paring edges and small pieces of leather or is sometimes referred to as a skiving machine. German made fortuna is one of the best known types of the machine.

(ii) **Board Sheer:** Guillotine and board sheer are best known machines for cutting board and paper.

(iii) **Stamping Press:** This machine is useful for tilting and die stamping. A number of makes and models of the press are available.

(iv) **Lying Press:** This press is primarily used for backing of a book.

(v) **Finishing Press:** Machine is used for tilting and tooling of a book spine.

Pressing Equipments

(i) **Press Boards:** Rectangular size with smooth surface, made of quality wood or plywood, are press boards used for fattenning documents and books while they are

Equipments and Tools

- Heavy Equipments
  - 1 Leather Paring Machine
  - 2 Board Sheers
    - (i) Guillotine
    - (ii) Board and paper cutter
  - 3 Presses
    - (i) Standing Press
    - (ii) French "
    - (iii) Stamping "
    - (iv) Job Cutter
    - (v) Lying Press
    - (vi) Finishing "
  - 4 Pressing Equipments
    - (i) Press Boards
    - (ii) Brass bound board
    - (iii) Guilding boards
    - (iv) Zinc sheets
  - 5 Glue or adhesive softening equipments
    - (i) Glue pot or milk boiler
- Light Equipments
  - Glueing and Pasting Equipments
    - (i) Glue pot
    - (ii) Past bowls
    - (iii) Paste brushes
    - (iv) Glue brushes
  - Tools
    - (i) Tools for paring leather
      - (a) French paring knife
      - (b) Spoke share
      - (c) German or English paring knife
      - (d) Skife
      - (e) Lithographic stone
    - (ii) Measuring tools
      - (a) Dividers
      - (b) Triangles
      - (c) T-Square
      - (d) Right-angle square
      - (e) Rulers
    - (iii) General Tools
      - (a) Sciessors
      - (b) Band nipper
      - (c) Bench knife
      - (d) Mat knife
      - (e) Spatulas
      - (f) weights (lead, brass, and brick)
      - (g) Bone folders
      - (h) Backing hammer
      - (i) Burnishers

pressed in pressing equipments. They are made of different sizes.

(ii) **Brass bound boards:** Press boards with addition of a protruding brass edge about 1/16 inch deep are known as brass bound boards. These are required for production of case binding.

(iii) **Gilding boards:** These boards are used for treatment of the edges of the books, particularly fore edges. They are very similar to backing boards.

(iv) **Glue pot:** It is thermostatically controlled electric pot for preparing glue. It maintains the temperature between 60°-66°C. A double boiler or milkboiler serves as a satisfactory substitute, but it should be used in conjunction with a cooking thermometer such as that used in making candy. It is important not to overheat glue. Fresh made glue serves better.

(v) **Glue brushes:** brushes made of pig bristles, round in shape and wire bound are considered the best brushes for applying glue. They are sold in numbers. A No. 2 is 1/4 inch in diameter while No. 4-3/8".

General Tools

(i) **Bone folder:** These are fashioned in many shapes and sizes and have many uses. A large one is used for folding of paper, smoothering slight wrinkles when gluing and pasting. The small folders are useful in setting headcaps, turning corners and drawing of lines to be tooled.

(ii) **Burnishers:** Like most of small tools, burnishers are also made in various shapes and sizes. The two types of burnishers, agate and metal are very common in use. Agate burnishers are used for polishing of book edges while metal one for building up German tube, ironing out small wrinkles and polishing leather. The metal burnishers are used hot.

(iii) **Backing hammer:** Hammer is used for backing of a book or vol. Hammer with a rounded head is useful for this purpose. These hammers are made in variety of shapes, sizes and weights. Sometimes for backing of small books, a hammer made of wood is very convenient. Selection of hammer is made according to work to be performed usually in the workshop.

Heavy Equipments

Fig. 5.21: Paring Machine

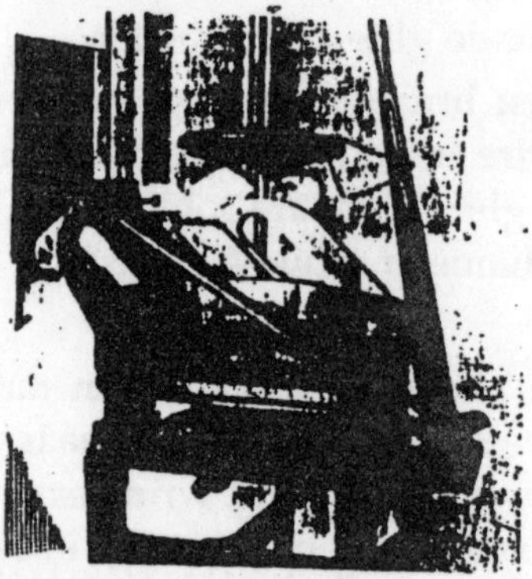

Fig. 5.22: Guillotine

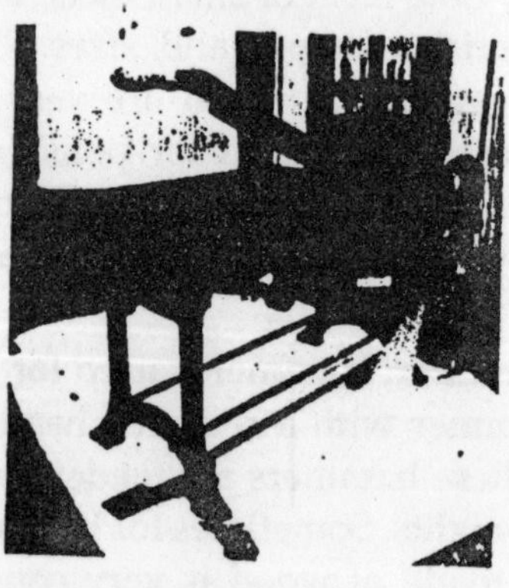

Fig. 5.23: Board Cutter

Presses of Different Type

Fig. 5.24: Standing Press

Fig. 5.25: French Press

Fig. 5.26: Stamping Press

Fig. 5.27: Job Backer

Presses and Pressing Equipments

Fig. 5.28: (i) Lying Press. (ii) German Backing Press. (iii) Finishing Press

Fig. 5.29: Pressing Equipments (i) Pressboards.
(ii) Brass-bound boards, (iii) Gilding boards (iv) Zinc sheets

Glueing and Pasting Equipments

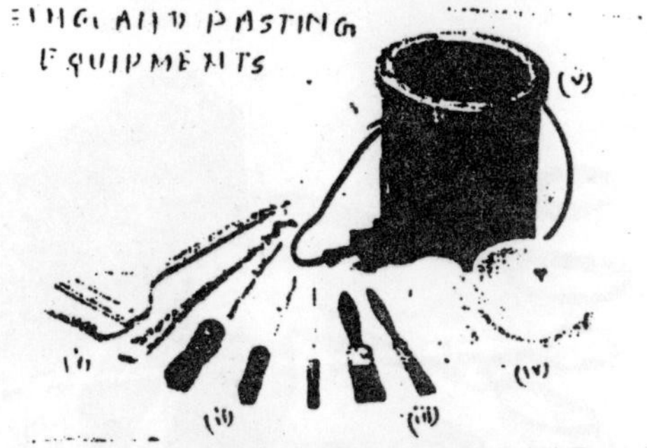

Fig. 5.30: (i) Japanese brushes, (ii) Glue brushes, (iii) Paste brushes, (iv) Paste bowls, (v) Glue pot.

Tools for Paring Leather

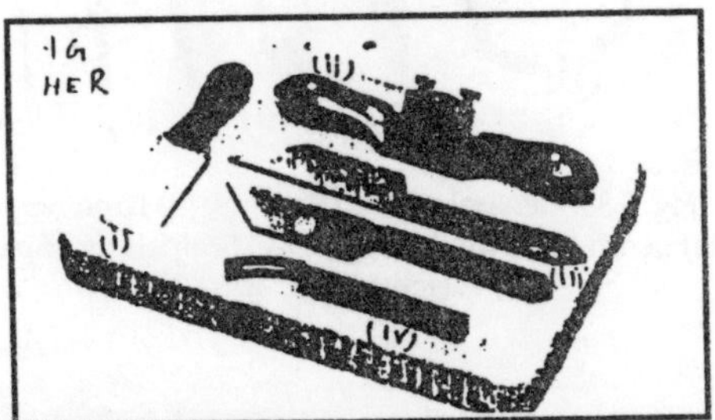

Fig. 5.31: (i) French paring knife, (ii) Spoke shave, (iii) German or English Paring knife (right and left handed) (iv) Skife, (v) Lithographic stone.

General Tools

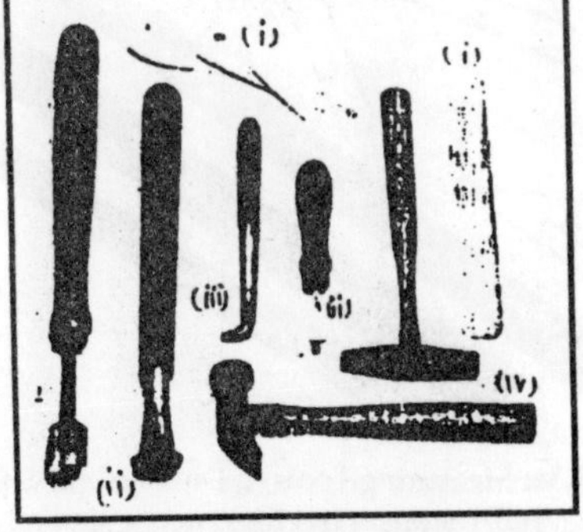

Fig. 5.32: (i) Bonefolder, (ii) Metal burnishers, (iii) Agate burnishers, (iv) Backing hammers

General Tools

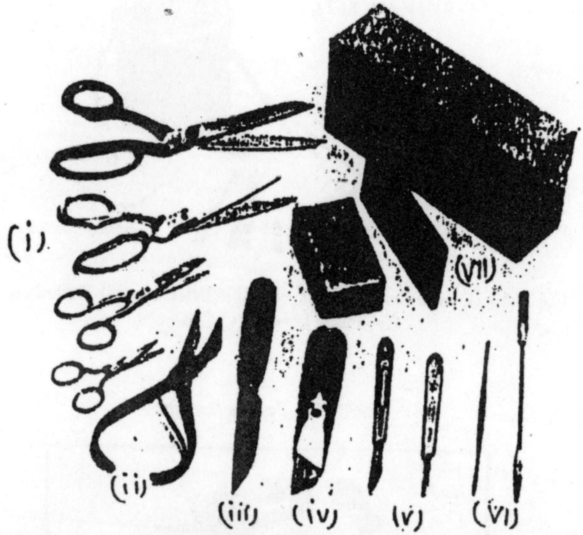

Fig. 5.33: General tools (i) Scissors, (ii) Band nippers, (iii) Bench knife (iv) Mat knife, (v) Scalpels, (vi) Spatulas, (vii) Weights (lead, brass, brick)

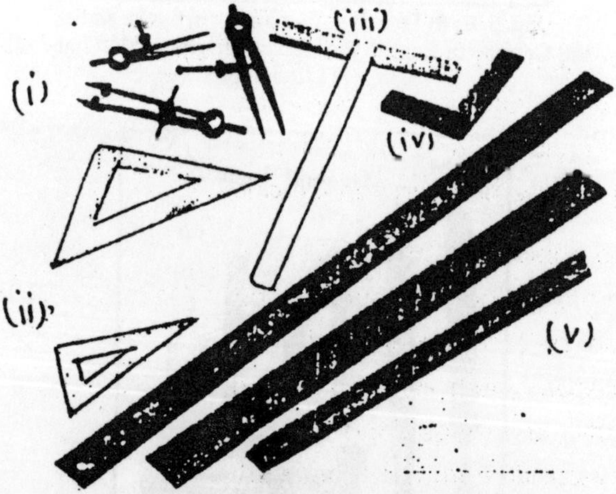

Fig. 5.34: Measuring Tools (i) Dividers, (ii) Triangles, (iii) T-square, (iv) Right-angle square, (v) Rulers and straightedges

A Ten Point Formula for Examination of Binding

1971, Edition of Library Binding Manual had suggested a 10 point formula to examine not only the fresh received bound books from the binary but also when the rebound book is worn out and is to be discarded. This practice will ensure whether or not the library has received its expenditure for the purpose worth.

1. General Appearance

For general appearance, a bound book is examined from sides, top, bottom and from closed back, to ensure.

(i) whether the cover is suitably attached at the spine. A too tight or too loose cover is considered unfair for.

(ii) Whether the cover proper size for the book? The cover, projected so far beyond the body of the book as to appear oversize and cut so close to the book itself as to give an over trimmed appearance, is not considered proper size.

Besides, see whether, the joints (grooves) straight and near, boards are of proper thickness for the book, cover free from signs of excessive warping and edges trimmed nearly.

(2) Open Cover (Back and Front)

Examine whether boards are cut neatly and uniformly and joints are flexible? Remember joints should be not too stiff.

(3) Other Covers

Examine whether cloth of proper weight, cut precisely and corners are neat. Also examine whether adhesion of covering material, cloth, paper or leather to board is adequate at all points?

(4) End Paper

While examining end paper, ensure whether they are properly constructed, strong, neutral colour, neatly cut and properly pasted.

All the end papers are consisted of three parts. (i) Outward end leaf or a pasted down which is firmly and evenly pasted down to become the cover lining. (ii) At least two free fly leaves, and (iii) Reinforcing fabric.

(5) Sewing

Examine to see whether the sections are of proper thickness, open flexibility, book tightly sewed and neat?

(6) Lettering

Examine whether lettering is straight, legible, centered, deeply and uniformly impressed and water proofed of a clear materials?

(7) Turn to Title Page

Examine whether title and instructions, if any, have been accurately followed in lettering spine of a book?

(8) Examine whether special instruction concerning the use of colour material and handling of maps or plates have been followed?. Whether the binder has affixed on the invoice, unless other wise noted, his warranty that the binding is LBI standard?

(9) Flat Examination

Lay the volume flat on a working table and starting from beginning and last sections. slowly and carely open and press down a few pages at a time progressing towards middle of the bound book or volume and see

(i) Whether the margins are adeque, paper is moderately stiff and sections scored?

(ii) Whether the book opens satisfactorily looking into the nature of the paper and the fact that the book is newly rebound?

(10) Take the closed bound book, slowly open covers all the back and fanout the pages from the middle, so that sections separate themselves naturally, and observe whether the sections are uniform and proper thickness and back flexible. The examination is applicable for machine sewed books etc.

This ten point formula is a technical guide to examine the correctness of freshly bound or newly rebound books and vols. However, careful inspection should not be confined to just one vol or two in each lot to many books as possible and by different staff members. The books and vols in which any mistake is found, are returned to the binding section without any loss of time for correction. A perfectly bound book or vol., however, deserves a appreciation to the binder.

Adhesives

Adhesives are very essential for reinforcement of books, Adhesives, lining and relining of paintings. In the synthetic world of today, there is flood of Adhesives. They are being used today in every sphere of life, including shoe industry and modern furniture industry for packing materials and joining of broken pieces etc. But the question is what is adhesive? How it was discovered and when? What is the composition of adhesive and characteristic requirement for restoring archives and other artifacts?

Actually, nature has provided a type of force, attractive or repulsive to every material. This basic force is working behind the adhesives. The common concept about Adhesives is that they help in joining two layers of materials technically known adherends. Every material displays the property of adherence at some specific point. For example, when two wet glass plates are kept together, one over the other, they require more force in their separation than they are dry. The reason is, when plates are wet, a force generated by water layer acts as an adhering force between the plates. Another example is that not only metals even stone at molten state works as adhesive. The adhering property displayed by these materials is more stronger than common Adhesives we are using in offices, Archives and Museums. Defining Adhesives, Amal Chandra Das writes that it is a class of materials which play a very vital role in production, reinforcement, repairing, restoration, and binding of paper based documents. Arun Ghose defines Adhesives, as an intermediary that acts as an agent for jointing two or more often independent materials.

The Adhesives which are available today, they vary in their composition. But the basic ingredients of traditional Adhesives (excluding thermoplastics) are the binding agents (in general solid) and the liquid medium. The emulsion of these two ingredients works as an adhesive.

The first Adhesive, when, where, how and by whom was discovered is the subject of research. But it seems that the use of materials displaying bonding character, specially for holding two or more ingredients together had started in India more than 17,000 years back for preparing pigments. It is evident from the fastness of colours on rock paintings of Bhim Baithaka (India) which are still bright without showing any sign of dullness even after age.

The Materials Showing Binding Property

Number of materials may be there displaying adhering property but some of them which have extensively used in the field of Archive and Museums are classified in four groups, making their origin a base.

Table 5.2 : Classification of Adhesives

	Origin	*Materials*
1.	Vegetable	Carbohydrate (starch & dextrine), Gum Arabic, Proteins (Soya bean), Natural Resins, colophony, Balsam, Natural rubber.
2.	Animal	Animal glue, Casein, Albumen, Bees wax, Shellac.
3.	Mineral	Silicate, Phosphate, Paraffin, Amber (Mineral resin), Bitumen (asphalt), Magnesia.
4.	Synthetic (Thermo Plastics)	(i) Carboxy methyl cellulose, Cellulose Nitrate, methyl cellulose, Cellulose Triacetate and cellulosics.
		(ii) Poly vinyl Acetate, Poly Vinyl Alcohol and Vinyl Polymers.
		(iii) Silicons, Vinyl Chloride, Poly sulphide, Ethylenc, synthetic rubber.

The Medium

The commonly used mediums for Adhesives other than thermoplastics are water and organic solvents.

Classification of Adhesives

Depending upon the nature of adhesives and their setting, property they have been classified in three following categories:

(i) Adhesives setting by evaporation/loss of solvents without loss of temperature.

(ii) Adhesives setting due to change of temperature.

(iii) Adhesives setting due to chemical reaction.

First Category Adhesives

(i) Aqua based: This category of adhesives generally have water or solvent base. They set by losing solvent, starch (maida) CMC and glue are the commonly deployed aqua based adhesives of this category. Besides, dextrin, methyl cellulose, poly vinyl alcohol, egg albumin and blood albumin also belong to this group. Because of the development of emulsion polymerisation technique many other synthetic materials such as poly vinyl chloride, neoprene rubber, poly vinyl acetate and poly acrylics have also joined the group of water based adhesives.

The adherend surface, permeable to water or water vapours are suitable for this category of Adhesives. Paper, wood and canvas which are used for reinforcement of manuscripts and archives and for lining and relining of paintings respond properly with water based adhesives emulsions. But permeation by these three materials is always accompanied by swelling when wet and shrinking after drying. Therefore, the moisture sensitivity of these materials needs prechecking before this category of adhesives and deployed.

(ii) Solvent Based: The adhesives are almost quicker in setting than water based adhesives and pronounce radiant results. Poly vinyl cellulose, ploy vinyl alcohol, ploy acrylates, ploy metha crylates, rosin esters, Nitto cellulose, neoprene rubbers and recent formulation BEVA 371 are adhesives belonging to this group. The solvent based adhesives like plexitol do not suffer from the disadvantages of the glue based mixtures and impart sufficient shearing force and offers low pull away possibilities.

This category of adhesives do not require heat to activate, set without loss of heat and are much dependable.

Second Category Adhesives

Hot melt adhesives called thermo-plastics are covered in this category of adhesives. No any solvent medium is involved for making an working adhesive of this category but they are made liquid by heating and applied. They set by cooling.

Acetate foil, waxes and wax-resin combinations being penetrative in nature permeates the surface of adherends.

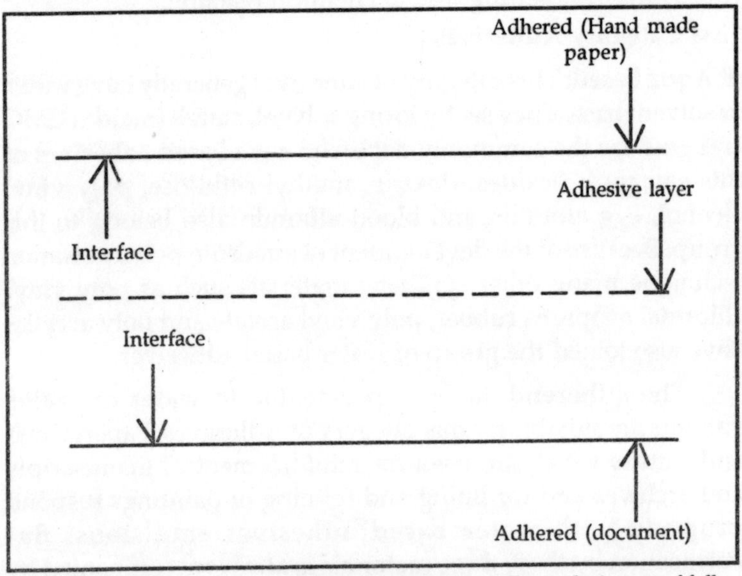

Fig. 5.35 : Showing elements involved in joining two adherends, in case of full-pasting a document.

Third Category Adhesives

The adhesives of this category are generally available in liquid state and made to set by addition of another liquid inducing a reaction between. Because of reaction no heat or little heat is evolved. The advantage of adhesives are that they neither shrink nor make interference in refractive index.

How an Adhesive Works

For joining two materials together, usually five elements are involved: See Fig. 5.35.

(i) Adherend
(ii) Interface
(iii) Adhesive
(iv) Interface
(v) Adherend

In case of full-pasting, handmade paper or any suitable paper and document are adherends and starch (maida) paste is adhesive.

Every adhesive has a capacity to form a film on drying and setting which displays a strength. Undoubtedly, both the adherends also have a physical strength. Keeping in view the strength of paper adherend, adhesive film and adherend as document or painting following three cases are possible.

(i) The strength of paper may be stronger than the paper of document but compatible with adhesive. In this case after setting, the reinforced document will curl down paper adherend.

(ii) The strength of adhesive film may be stronger than the strength of paper adherends. In this case the resultant sheering force during drying may tear away the paper.

(iii) If the interfaces of the adherends are not suitable permeating adhesive or the adhesive is weaker than the required strength for adherends, in this case the setting will result in loose joining. This character of adhesives and adherends is termed as compatibility. For proper joining, the compatibility of adhesive with the adherends is very essential.

Properties of Adhesives

Viscosity, reversibility, toxicity are few properties of adhesives which need hardly any explanation. But shrinkage, cold flow, permeability to toxic vapours and moisture, bonding strength, flexibility, penetration or impregnation and softening point need a bit explanation for clear understanding.

Shrinkage

Films of certain adhesives, after setting, shrinks. The shrinkage sometimes causes the adhesive to pull away from the adherend, if it is not sufficiently penetrated the layer. If adhesive after setting shows shrinkage, it causes to develop a stress and strain on the fibres of paper. The tension so developed hardens the surface and with the passage of time results in damaging it.

Gum adhesives are the best examples showing this property. This is the reason that gum adhesives are not preferred for restoration of documents.

Cold Flow

Some adhesives under tension, ooze through or go out of the adherend layer. This property is termed as cold flow. Because of it, adherend suffers of adhesive starvation resulting in insufficient bonding.

A full-pasted document, if pressed in a *dab* press, before proper setting of the adhesive (starch paste), the adhesive will come out of the edges and proper bonding of adherends will not take place. As such, documents so repaired are pressed for removal of wrinkles when adhesive is properly set and not prior to that.

Permeability to Moisture and Toxic Vapour

Some adhesives are such that they show affinity to moisture. Means they absorb water vapours from the surrounding environment and become wet. Glue, starch and dextrins belong to this category of adhesives. This property, is conducive for promoting fungus growth. However, mould growth can be checked by good house keeping and adopting preventive measures.

Certain adhesives are reactive to toxic vapours and gases like sulpher dioxide and carbon dioxide etc. present in the atmosphere. This property of adhesives, endangers the life of documents imparting acidity to them.

Flexibility

Certain adhesives improve the flexibility of a document while some decrease it. Improvement of flexibility is a regardable property in restoration of documentary heritage.

Choice of Adhesives for Conservation in the Tropics

For quality reinforcement of documentary heritage, lining and relining of paintings, an adhesive should have following characteristics:

(i) An adhesive, should never develop wrinkles or shrink after setting.

(ii) It should not be hydrophilic and reactive to the environment gases.

(iii) It should be characterized by forming stable and flexible film.

(iv) Stability, durability and reversibility are the desired characteristics a quality adhesive.

(v) It should be colourless and not change its colour after setting.

(vi) It should be free from acidity.

Adhesives (thick and thin) prepared from wheat flour (Maida), glue and Carboxy Methyl Cellulose (CMC) are normally preferred in Indian sub-continent for restoration and binding of Archivo-Library Wealth.

Care and Salvaging of Fire and Water Damaged Records

Fire and Water are the potential destroyers of documentary wealth. The extent of damage and problem caused by these two are mostly common with the exception that in the former case mass of paper suffers with varying degree of charring while in latter most of Archives suffer with running of ink and colour and with other problems. In the post occurrence of both the incidents, if records and other documentary materials are not handled with care, they may attacked by various damagial agencies. As such handling of problems arisen out of both the cases, need a lot of preseverance yet setting up of a system preventing damage, quick salvaging and restoring the bulk requires serious attention and meticulous planning.

Salvaging Fire Damaged Documents

A fire accident in any repository, if controlled successfully, the lot of fire hit records may be found in four categories, the completely burnt, the partially burnt, the partially charred and the slightly charred.

Salvaging completely burnt records is a futile exercise as they are reduced to ash. However, partially burnt records (which have converted into charcoal) can be restored, if attended by experts and the skilled personnel, through stabilising such documents using poly vinyl acetate and subsequently deciphering them with the aid of ultra-violet radiations. The text may be brought out by photo-graphing such documents. This category of records require immediate attention for treatment and examination without any loss of time. The action delayed, generally results in loss of information.

Charred papers being fragile need separation and placing each sheet in a separate cover for safe storage. Any amount of pressure to such papers, may shatter them into pieces. The covers containing charred papers are placed flat on a shelf or transfer box.

Deciphering of badly charred documents to bring into light the written contents of the text is followed by photocopying which requires the help of infra-red radiation technique. If instant deciphering is not possible and the transfer of such records to the expert agency is also not feasible, they should be kept in the repository as separate collection. Every effort, however, should be made to get the lot deciphered and photographed as a matter of urgency.

Partially charred documents, according to their condition, however, are treated suitably and restored.

Salvaging Water Hit Records

Disasters (Flood, hurricanes, use of water in fire fighting) improper storage and plumbing failures are the prime reasons of damaging books and other record materials by water. Besides, serious roof leaks and leaks from defective plumbing and seepage, entry of showers, through windows and ventilators also results in wetting the material and causing serious damages.

The damage caused by the water, bring about loss in mechanical strength of paper, loss of sizing materials, running of delegible ink and growth of fungus.

Experimental studies have revealed the fact that the strength of water soaked, paper is considerably lower down than the unsoaked paper. Consequently, the handling of wet documents causes damage to paper.

Sizing materials, such as glue, gelatin and rosin, if paper is surface sized, dissolve in water and smudges often leading blocking of books and documents.

Slow drying of wet records consequently favours fungus and moulds growth especially if paper is sized with glue and starch.

For salvaging wet records of all kinds, first of all such records should be transferred to an airy room. The books, volumes, files and bundles instead placing flat are stored vertically keeping spine on the top facilitating rinsing of water. Subsequently, according to the bulk of wet records and their nature of damage, priority treatment is fixed and work is taken up as under:

(i) Cloth used for tying of *bastas*, if any, is removed and kept aside.

(ii) Wet sheets of files are carefully lifted from one another and placed individually between white blotters. The code number and page number in sequential order is marked over the sheet/blotter, as per convenience to identify the sheets and their arrangement in original order, after drying. The room is made airy using fans, air circulators, exhaust fans, etc. in summer season while in winter temperature of room is raised using heaters, heat convectors, etc. No wet sheet is placed near by any source of heat directly for drying.

(iii) According to the capacity of the room, maximum number of files are so treated before they dry.

(iv) Now, the sheets are allowed to dry. If need be blotter should be replaced by new lot and marked with the same number allotted to old blotter.

The books, made of good paper, however, do not suffer seriously because of wetting. The sheets of the books are interleaved with quality blotting paper and allowed to dry under pressure. If necessary, the blotting papers are replaced by fresh lot. Using the process all such books and volumes are treated.

Sheets and covers of thoroughly soaked books are separated, placed individually between blotting paper and dried under pressure. If required, the dried sheets are resized to improve physical strength and smoothering surface. The book containing art papers, must be given proper attention, as sheets after a time stick together and becomes difficult to separate without damage.

The blocked sheets of a book or volume may be treated with the jets of hot stream (100°C) which helps in destroying bacteria causing putrefaction and also in separation of blocked sheets. So dried sheets of files and books are arranged in original order and interleaved with thymolised paper, to prevent any mould or fungus growth.

Protein or starch sized papers, if dry after wetting, stick to each other making a solid mass. Such blocked sheets are difficult to separate using ordinary means without damage. Separation of such sheets, however, require the use of enzymes for releasing and separating blocked sheets easily.

Animal glues are composed largely of protein. They are sometimes added to starch based adhesives in varying amounts. Sudhir Kumar Das, who worked with the problem, suggests the use of 0.5% W/V Trypsin, which is capable to breakdown any proteinous adhesive film likely to be encountered without first determining its composition. The enzyme is applied at optimum temperature; say about 40°C and pH 7.4. The use of protein enzyme solution of concentration 0.05% W/V at temperature 40°C and pH 7.45 has also demonstrated positive result. Its use, however, prompts a few words of caution. Coloured inks and paints soluble in water should be fixed with 5% solutions of cellulose acetate in acetone before the document is subjected for enzyme treatment. Gold leaf attached to a document with egg albumen are readily attacked by enzymes hence should not be treated with the solution.

Starch sized blocked papers are separated with care using 5% medicinal grade diastase in water at room temperature say 20°C, without any adverse effect on paper and ink. The starch pastes consisted of amylose and amylopectin (both are polymers of glucose) in proportion 25% and 75% respectively, are broken into microscopic particles and softened using lamylase, making removal easier. The optimum temperature and pH required for application of lamylase are 37° to 39°C. and 6.95% respectively. The upper limit of concentration of enzyme suitable for appropriate action, is 0.59% W./V. The rate of reaction depends on the rate of penetration of reagent into the paper support.

The excess enzyme is removed by washing in two changes, in ordinary tap water at a temperature of 20°C followed by a hot rinse in tap water having temperature 50°C.

Above methods, for separation of protein or starch based adhesives, should only be used if ink is not soluble in water.

Method of Application Involves Following Steps

(i) Determine the kind of adhesive whether it is starch or protein or mixture of both. It will help in selection of appropriate enzyme.

(ii) Store enzyme bottles in polythene bags with some silica gel in the freezer part of refrigerator.

(iii) Make a stock of buffer solution. Remember buffer solution has a limited self life should be stored in refrigerator.

(iv) Warm 900 ml. of non-logic water to 40°C and add 100 ml. buffer solution in it.

(v) Place dish in hot water sink, add warmed dilute buffer. Ensure the temperature of dish is 37°C. (using thermometer) as enzymes deteriorate over 40°C and are ineffective in the cold.

(vi) It starch is to be removed, add one gm lamylase to buffer and stir it gently.

(vii) Place document between wet strength paper and immerse in enzyme.

(viii) Agitate dish occasionally and keep on the temperature.

(ix) Reduction of starch takes a time ranging from 30 minutes to 4 hours. For testing the presence of starch, put blotting paper against the document and then dip it in dilute iodine solution. A purple colour indicates the presence of starch.

(x) In case of removing protein, make fresh warm dilute buffer solution as before and add 0.1 gm. of protease maintaining temperature 30°C.

(xi) When all traces of glue and paste are removed, rinse, deacidity and resize the document under treatment.

Principles of Documentary Restoration

The restoration of paper based documentary wealth is not mere the application of paper and paste. The purpose of restoration may defeat its own objective, if the true nature of document, material and adhesive is not understood properly. It is the character of archive or record that determines the nature of their treatment or the course of action to be taken to improve its physical character. Even slightest error in selection of right material, appropriate technique with which document is to be subjected may bring about irrecoverable loss to a document which is undesirable and unwanted. No process of repair should be allowed to diminish, remove, falsify or obscure, any way the documents value as evidence. To eschew all unwanted happenings with document, and to restore it ensuring lasting longevity, an eight-point code of ethics, is available now to carry out restorative conservation of archivo-library materials. This code is known as parameters or principles or rules of restoration. These parameters have evolved because if the experience of the conservationists world over. In one way this code is a guiding force to achieve faultless and desired reinforcement work. The parameters advocate that.

(1) **The written contents of document should not be marred or impaired because of their chemical treatment and restoration.** Decayed and fragile documents are subjected for chemical treatment and restoration to keep the recorded information alive. The document would have never been subjected for the purpose if the importance of information had not been realized. As such, if written characters are even slightly deleted or their hue happens to fade, the purpose of work is defeated. To ensure that nothing such would happen during chemical treatment (deacidification) and repair, the behaviour of ink, against the solution to be used for chemical treatment and adhesive to be employed for reinforcement are tested before the document is subjected for. It is not advisable at all to deacidify a document written with water soluble ink in any solution prepared in aqueous medium. Likewise the reinforcement of documents written with deligible inks should be avoided employing adhesive having aqueous base.

(2) **The process employed for coating and reinforcing a document should be reversible.** The concept of reversibility is a very vague term but in the profession of restoration it implies that the adhesive and materials employed for reinforcement should be such that they may be reversed (removed) as and when necessity arises without bringing about any adverse effect over the support and the written characters. Necessity of reversing the process arises, if it is noted that the document reinforced earlier is suffering or decaying fast because of

contents of the adhesive or the material employed earlier for reinforcement or because of the contents of a document. To save documents from ill effect, removal of materials employed earlier for repair and chemical treatment becomes essential as without reversing the process the cause of decay and imbrittlement cannot be removed from the document and without it, the expectation for stability, permanence and durability of re-reinforced document is futile. Reversibility has become necessity today because of availability of ordinary laminators, unstable materials for lamination at cheaper rates and synthetic adhesives. But reversibility of laminated sheet through ordinary cheap materials is a dream. The fate of document laminated with such materials is shield permanently. At any stage, the document cannot be removed from the material employed in lamination. If any how, it is managed that would not leave the document unharmed. Moreover, the cheaper quality of laminating materials give rise to noxious fumes in the stack areas, adversely affecting both, the documents and the people working in the area. On the contrary the stable materials confer permanence and durability if laminated document is stored in monitored repositories.

(3) **The process should impart maximum strength at minimum cost without hampering physical strength of document.** Repair of documentary heritage is a very costly affair which can be visualized from the cost involvement in different. Repair techniques. The reinforcement cost of an A4 size document, employing lamination, tissue repair and full-pasting comes about Rs. 20, Rs. 8 and Rs. 4 respectively. Lamination cost of octavo size sheet is approximately Rs. 12. If every sheet of a file containing 100 document is laminated, the total expenditure involve would be not less than Rs. 2000 excluding binding charges. Lamination being good looking, clean, giving radiant result and being less cumbersome, once restorer masters the technique, always, prefers it comparatively to other techniques. But any organization of developing and under-developed country cannot afford this cost. Archival organization even in developed countries are facing

problem in running their restoration work because of shortage of funds. Then question is what to do? The answer lies in the fact of being economic. One should be very selective in technique of reinforcement. Selection of technique should be made according to the condition of a document. If a document can be reinforced with full pasting, tissue or any other technique that should not be laminated. The latter should be employed only when no alternative to restore the document is available. However, lamination man be compared with life saving drives to save the endangered records. Such records if not reinforced within a stipulated period, they may turn into dust. If any such eventuality is foreseen, the mass repair work is the only solution to save them and that can be done by the machine lamination only waving the cost factor.

Typed documents signed by the authorities in delegible inks are generally laminated, to restore their strength. Such documents can, however, be reinforced at cheaper cost, if proper attention is paid. It can be achieved by fixing ink first and then repairing. A 5-10% solution of acetate foil in acetone if pasted over the characters, it fixes the ink against water and facilitate reinforcement even using adhesive prepared in aqueous medium.

(4) **Legibility of document should not be affected.** It is a well established fact that written contents of a document are valuable. They should never be allowed to fade, diminish partially or fully, because of chemical treatment and repair. It requires, the expert selection of material and technique for repair, specially in case of documents whose ink has already become dull. Though in general, a colourless adhesive with adequate bonding strength and covering materials almost colourless are recommended for reinforcement, yet in case of document with dull-writing, chiffon repair and chiffon lamination are not suitable. Chiffon to some extent obstruct the transparency because of its texture and colour and hence legibility.

(5) **The reinforcement must never become temporary and never can become so provided that it obscures nothing of the original.** This statement seems abit unvogue to a

non-conservationist but it is required. Not only the written characters but figures, drawings, motifs, colours and even materials and their fabric sometimes convey very important information. Conservators should believe that all the features of a record, even those which are unwritten are to be preserved for their information value. Let me make it more clear through an example. A bound book with deer skin belonging to 15th century A.D. decorated with golden and silver ink employing unique motifs style is received in the conservation laboratory for reinforcement in India. For an ordinary restorer, the writing ink and artistic decoration including material (deer skin) are meaningless. He will dismental the binding, throw the material in a dustbin ruthlessly, presuming them useless and rebound the book with new material after necessary chemical treatment and repair of torn sheets. The look of the newly bound book would be totally different from the original. Obviously, the originality as well as the antiquarian value of the book is totally changed. One may ask as to how this change in originality is affecting the scholarly world. The answer can be percepted from the facts mentioned below. In the subsequent period of completion of repair of aforementioned book, presume that one article published in a journal claims that

(i) Deer skin was never employed for binding of book and vols.

(ii) The golden and silver colours were employed for decorative work first time in 17th century.

(iii) The specific motifs and style (employed over the book of 15th century referred to above) were employed first time in Europe.

If no other source regarding material, colour, motif and style which were employed in 15th century volume (referred to above) is available, the three facts mentioned above can not be challenged though they are not correct and the truth lies on the facts:

(i) Deer skin was employed for book binding even in 15th A.D.

(ii) The use of Gold and silver colours was prevalent in 15th century.

(iii) The motif and style being claimed of European origin was being employed in Asia in 15th century.

The source to corborate all these facts was the old book of 15th century A.D. which were unwritten. The importance of retaining originality of a book and the importance of unwritten source of information can be visualized from this example. As such not only the books and volumes but such sources also which are found in old manuscripts should be retained during the process of reinforcement. Even, the size of document should not be reduced and the unwritten papers, available in old volumes, should not be removed and destroyed at the time of repair. If they are in good condition, they should be made part of the rebound volumes/book etc. If, necessity of destroying them arises, the matter should be brought into the notice of higher authorities, for decision prior to their removal and destruction.

The question is as to how the originality of a bound book or volume and its antiquarian value can be preserved when the board and covering materials are damaged. It can be done very simply. First, before dismantling the binding, take the photograph of front and back of the volume. Ensuring the exposure is successful, dismental the volume but during the process every motif, design and written portion should be kept intact separately to the maximum possible extent. Rebind the volume using new quality material. Clean material and old motifs, design and letters, if necessary. Now with the help of photograph, set them in original pattern employing quality fixatives. This will survive the original make-up of a bound book or vol.

Every uncommon artistic decoration on the old book and volume is rare and may be an unique source of information. If this thought works as passion of an artist and technocrat, he may save the unique source of information from permanent loss. This is the dedication which requires in archival profession.

(6) **The Process of repair and binding should be durable and permanent.** The record is to be preserved for eternity and the purpose can only be served if the material selected for restoration of documentary wealth pronounces adequate strength and conform the quality. If an acid free document is wrapped with acidic paper or cloth, the decay of document even after its repair will continue. A repaired document or volume having permanent and durable character is a product of sound materials and good craftsmanship. To add permanence and durability to a document or book or volume, because of their repair, the use of quality material is essential. An archival document is not a thing of frequent repair. The effort should be that the document repaired once may not need re-reinforced during century in normal condition. It can be achieved only if all the elements present in a document endangering it, have been removed prior to its reinforcement and the quality materials are employed for rehabilitation.

Fibrous mateials like paper, board and cloth and non-fibrous adhesives never be procured on cheaper basis but they should be only adopted for use if found suitable on test grounds. So far as paper, for restoration work, there is no match of handmade paper produced by KVIC. In India, it is known for its multi-purpose use i.e for full-pasting, guarding and gathering, filling up of get-in-paper, preparing end sheets and even for covering binding boards. Really, this paper, upholds, the Indian proverb "Ek Anaar Sau Bimar" in conservation.

Briefly speaking, the permanent and durable repair of records and books is the result of employing every material of quality tested by time.

(7) **The restoration of a document should bring about minimum change in bulk.** A restored document, gathers thickness and weight. Consequent upon, the weight and bulk of a book increases. Sometimes, bulky books having sheets more than 2000 or so, because of restoration of each sheet acquire more than double bulk and weight than original. The higher bulk and weight cause inconvenience in handling and storage. Because of it the book or volume, deteriorates with rapid speed which is undesirable in

conservation. As such, every effort should be made not to allow the bulk to increase enormously. In case a work gathering more bulk due to reinforcement that should be bound dividing into convenient parts.

(8) **While restoring the documents and books, aesthetic part should not be ignored.** There was time when most conservationists gave much weightage on the technical aspect to save the information and the aesthetic part was quitely ignored. But the latter too adequately be considered which is the source of attraction among the human beings. A thing of beauty is really a joy for ever and develops a bond of attraction between the two, loving and lover. No doubt, lover always comes from the group of living organisms but loving sometimes may be a natural object or a piece of art or any other thing including document. This bond of love keeps the two psychologically attached for a long enduring charm which subsequently give rise to other things but in case of a document or book, if it is nicely and beautifully worked, will induce the charm for study. If it is a piece of ugly patches, rough and dirty, the charm of going through the text or document is considerably suppressed. Keeping all these facts in view, every effort should be made to accomplish any repair pronouncing beauty. For the purpose, not only the work should be neat and clean but the paper and other materials for reinforcement should match the colour of document. But any dispute, between beautification of Book or Vol. and preservation of information, the latter should be given top priority.

In the last, but not the least, nothing should be allowed to happen during the process of treatment and repair that hampers document adversely.

REFERENCES

Aggarwal, O.P., Suryavanshi, D.G. and Gupta, C.B., "Effect of Movilith on Fibrous Materials", *Conservation of Cultural Property in India*, Vol. XVI & XVII, 1983 and 1984.

Chakravorty, S., "A Review of the Lamination Process", *Indian Archives*, Vol. 1, No. 4, 1954.

Das, Amal Chandra, "Adhesives", *Conservation of Cultural Property in India,* Vol. XVII, 1983, 1984, p. 90.

Das, Sudhir Kumar, "Repair & Preservation of Water Damaged Documents", *Conservation of Cultural Property in India,* Vol. VII, 1979, p. 43.

Ghose Arun, "Note on the Classification of Choice of Adhesives", *Conservation of Cultural Property in India,* Vol. XV, 1993, p. 18.

Goel O.P., "Repair of Documents with Cellulose Acetate Foils on Small Scale". *The Indian Archives,* Vol. No. 2, 1953, p. 162.

Grant Julius, "Analytical Methods in the Dating of Books and Documents," *The Indian Archives,* Vol. 1, 1947, p. 30.

IS: 3050-1950-binding.

Kathpalia, Y.P., *Conservation & Restoration of Archive Material,* pp. 78 to 145.

--------, "Restoration of Fragile Document", *Conservation of Cultural Property,* Vol. X, 1977, p. 33.

Lal, Avinashi, "Destructive and Non-destructive Methods of Analysis," *Conservation of Cultural Property in India,* Vol. XI, 1978, p. 54.

Lal, B.B., "Scientific Examination of Documents, Works of Art, Monuments, and other Cultural Property", *Conservation of Cultural Property in India,* Vol. XI, 1978, p. 21.

Prajapati, C.L., "Archivo-Library Materials, Their Enemies & Need of First Phase Conservation", p. 54.

---------, "Conservation of Manuscripts in Tawang Monastery", *Indian Archives,* 1992.

Prasad, S.N., "Some Observations on Conservation of Archival Material", *Conservation of Cultural Property in India,* Vol. VI, 1973, p. 11.

Scol & Tamas, "New Method of Determinating the Surface pH of Paper by Plane Glass Electrods", *The Indian Archives,* Vol. XXI, No. 2, 1972, p. 51.

Sharma, S.K., "Scientific Examination of Documents", *Conservation of Cultural Property in India,* Vol. XI, 1978, p. 31.

Smith, Herman, "Manuscript Repair in European Archives", *The Indian Archives,* Vol. II, Nos. 2-4, 1948, pp. 90 to 111.

---------, "Manuscript Repair in European Archives (France, Belgium and Netherlands)", *The Indian Archives,* Vol. III, Nos. 1-4, 1949, pp. 96 to 121.

Talwar, V.V., "Care of Water and Fire Damaged Records", *Conservation of Cultural Property in India,* Vol. X, 1977, p. 40.

6

PROBLEMS IN CONSERVATION AND SOLUTIONS

Onceptionally when modern National Archives of India came up as Imperial Record Department in 1891, no infrastructure of conservation was available. First time during the period of Samuel Charles Hill (1899-1903), the work on flattening of folded records, repairing of damaged documents using tracing paper and their placement in docket covers was introduced as sole conservation. Subsequently, in the tenure of Alwain Faber Scholfield (1911-12), the tracing paper was replaced by more enduring chiffon cloth.

The Indian conservation has faced various problems in the past. But, due to untiring efforts of custodial agencies, it could borne the fruits by finding suitable answers of the problems. The problems which merit are of various nature and belongs to every element of conservation; examination research and development, preservation, restoration and duplication. Mr. M.L. Gupta in this regards has stated that some problems relating to manuscripts are concerned with non-availability of materials, retaining of professionals in laboratories and workshops and the training needs. However, the problems Relating to Establishment of Effective Preservation and Restoration Programme in Archives and Libraries", could be categorized in two; problems of collective nature and problems of individual nature. The first one is further subdivided in four groups, (i) Lack of Archival Awareness and Technical Known

How; (ii) Shortage of Standard writing materials for creation and storage of documents and equipments; (iii) Diversity of climatic conditions; and (iv) Lack of skilled conservators and restorers. Nothing is wrong to say that the lack of incentives and better promotional avenues are considered today the great obstacle in development of conservation as science as well as arts with fast pacing.

This classification is not water-tight and exhaustive but its significance lies in the sense that it provides necessary base for studying the problems of various nature systematically.

(A) Problems of Collective Nature

Lack of Archival Awareness and Technical Know-how

The lack of archival sense among the people and also the lack of technical know-how to preserve and restore the wealth, have been the main causes behind to throw the heritage in appalling state. The resolution passed by the Historical Records Commission in 1942, gives the clear apprehension of this fact.

What necessary steps, to arose awareness among the people and to disseminate the technical know-how till 1940, were taken, is difficult to say but in the post independence period, the exhibitions, deputation of officials to professionally advance countries and publication of journal worked as main tools to sort out these problems and achieve tangible results.

In 1947, National Archives of India, during the period of Dr. S.N. Sen (1939-1949) then the Director of Archives (1945-1949) deputed a team of officials of this Department for training in the National Archives of Washington, U.S.A. This training gave impetus to the work of finding out solutions of various problems indigenously. A conservation research laboratory in the Department in 1940 had already been set up for this purpose. The attempt of 1947, to get the staff trained, was a judicious step to hammer the nail on the face of problems as without the practical knowledge of the field, it was not possible to run the work of laboratory and find out means and ways to cope up with day to day requirements. The practical knowledge of the field helped in building confidence among the officials to take up the work independently. Publication of journal, "The Indian

Archivist" (started in 1942) also helped in disseminating technical know-how to the people of custodial agencies.

Because of these efforts people soon understood the role of functional archival building and air-conditioning in promoting the cause of preservation and in turn, authorities of archives became assertive and instrumental to achieve them. Now, the situation is that not only National Archives but most of the Indian States have scientifically designed functional Archival Buildings. Those which are not having persistently trying for. The necessity of a monitored repository is well understood by many but money is a great problem to achieve that. With persistent efforts of Archives, the creating agencies are also endeavouring to set-up model record rooms, (see at Appendix VI) for housing records to preserve them scientifically till they are transferred to archives for their proper preservation.

Problems Relating to Diversity of Climate

Role of climate is pivotal in preservation of documentary heritage which can be visualised from coverage in Chapter 3. Climate of Indian sub-continent is not uniform. Some regions are marked by severe cold (North and North-east) while some by high humidity. Some parts of the country reel through scorching heat in summer (Central and Western Parts) while some regions are affected by the frequent variation of temperature because of hide and seek played by the sun and the clouds. Western part of the country is marked for an otherwise different phenomenon. In day time temperature in summer rises to 47°C but in nights comes down to 30-35°C. The frequent change of weather, as well as temperature brings about nothing but ravages to record materials of organic nature.

About the diversity of climatic conditions, it may be stated that the climate of earth has been continuously changing and atmosphere has been polluting since evolution of life. Thanks to natural ecology which has been maintaining balance. But fast industrial growth, race of material achievement and mindless deforesterisation all over the world have brought destructive changes in climate and atmosphere beyond the capacity of nature to balance. Consequently, not only mankind is suffering because of fatal diseases but priceless wealth of archives and libraries have also reached the brink of disaster.

The right answer to counter the problem is to have a pollution free and hygienic atmosphere. But it is far cry now in prevailing conditions. However, it is very essential to maintain these in Archives and Libraries where riches of cultural heritage rest. It can be achieved by going in for the air-conditioning and introducing air filtration system in muniment rooms. Stable indoor environment in a repository however, can be had by controlling humidity and temperature at optimum ranges.

National Archives of India though got approval of Government of India in 1947 to achieve this facility yet due to one reason and another, it was availed in 1970 only. Annex as well as old building of National Archives of India is availing the facility now.

Problem Relating to Shortage of Standard Writing, Covering, Reinforcement Materials and Equipments

Wrong choice of materials for creation of records (paper, ink, carbon-paper, ribbon-ink); covering materials (docket covers, carton boxed, board and cloth) and materials required for reinforcement of records instead delivering any good in prolonging the life in general happen to retard that. Besides, rough handling and poor storage also adversely affect the life of records. As life expectancy of an ill-born, poorly nurished child without special medical aid is always low, in the same way a document created using poor quality of paper and ink, uncared and poorly maintained is thought to be an object of short life.

The paucity of materials, technical know-how, is being felt since Archives as the Imperial Record Department started to roll on. However, facility of conservation research Laboratory and availability of trained staff, paved the way for stream lining the work of conservation in right direction evolving maximum of indigenous sources. Testing of covering paper in 1972, testing of cultural paper in 1978, and study of fungicides and insecticides in conservation Research Laboratory to know their suitability, are few examples, of this category. Because of this it could be possible to have quality paper for docket covers, paper for creation of records of permanent nature from indigenous sources.

The problem relating to sterilization of records, was solved by introducing commercial insect repellents and fumigants after testing their suitability in the laboratory. In 1939, the use of Naphthalene as insect repellent and fumigant had already been introduced. Subsequently, the fumigation techniques using Paradicholorobenzene, killoptera and vacuum system were adopted in 1941, 1949 and 1951 respectively. The adoption of these materials and techniques provided a suitable system of keeping records insects free. Introduction of thymol fumigation in 1955 gave a suitable technique to keep documents safe against the ravages of fungus and also to save those suffering from fungus infestation.

Though specifications for inks[+1] and ball point pen inks[+2] for creating permanent records are available, yet need is there to make concerted efforts by all creating agencies and Archives, to make them available. Discussing the matter with leading ink manufactures of the country and to pursue them for the purpose may bring about tangible results.

Other materials like carbon paper[+3], typewriter and ribbon ink[+4], stamp-pad ink also have their role in creation of records. If their quality is inferior, they may pose problems arising out of fading. Stamp on a document confers its authenticity. If it fades, may give rise to the legal problems, for authenticity of document. As such, to help the creating agencies, ensuring quality of matricides, from the view point of preservation, Indian standards on the subjects were got formulated.

A suitable record room in creating agencies is essential to preserve records before they are transferred to Archives. But, it is unfortunate that very few agencies have model record rooms. National Archives of India has been pursuing the importance of departmental record rooms, well equipped and well developed, for the purpose through lectures workshops and organising in service training programmes. Recently in 1999, the department prepared a layout plan for model record room and the same has been sent to the Department of Personnel for circulation all over the country.

In order to ensure effective conservation of National Documentary heritage, few standard like recommendations relating to primary elements in the designs of building for libraries (IS: 1553-1989), recommendations relating to primary elements in the design of building for archives (IS: 2663-1989) and standard specifications for quality binding IS: 3050-1965, were got formulated sharing platform with BIS.

Lack of Skilled Conservators and Restorers

Acquired technical know-how from abroad during the period of Dr. S.N. Sen (1919-49) when Archival profession in India was at its infancy and moving in lymlerinth, set direction of conservation in India. It not only paved the way for testing suitability of materials but also advanced to develop new techniques of reinforcement and storage. Solvent lamination, special full pasting and a technique of improving storage system of Palm Leaf Manuscripts are some new techniques developed in Indian conservationists.

Solvent lamination was developed by Mr. O.P. Goel in 1951. The latter two by the researcher himself in 1991 and in 1994 respectively. Besides, a paste from Sarcell HV and Sarcell MV for repair of document using tissue paper was formulated. An emulsion from CA powder was formulated in the CRL in 1996. The constituent of this emulsion are, however, the same which are present in cellulose acetate foil. The CA powder unfortunately is not locally available now.

Cater to the requirement of custodial agencies of trained personnel National Archives of India (then Imperial Record Department) had started in-service training scheme in 1928. Later on this training activity was upgraded and an Institute of Archival Training within the department was set up in 1976 which further re-christened as School of Archival Studies in 1980. Training courses of professional and sub-professional level for conservationists and restorators were introduced. One such course of sub-profession level in Repair of Records (now servicing and repair of records) was introduced in 1973. Subsequently a course in care and conservation of Books, Manuscripts and Archives of professional level was started in 1980. Both the courses are short term courses, each of eight week duration (six weeks now). Besides, various workshops on repair of records have been organized by the department to make the people aware about the reinforcement techniques for scientific mending of archives and other documentary heritage. The courses are so popular that the candidates not only from within the country but from abroad as well come for training.

The Conservation, no longer remained a secret of urban class but it set its feet in the remote hill areas of North-east (Tawang) and Islands of Majuly (Assam). The people of Tawang Monastery were trained in 1992 to make them enable to conserve their riches of collection. Satras people of Majuli Island were imparted on the spot guidance for scientific preservation of Sanchi Pat Manuscripts in 1999.

Problem in Conservation of Outsized Charts, Maps and Newspaper

Outsized maps, charts and newspaper pose a serious problem in their preservation and restoration. The suitable answer for proper storage of outsized records is their sectional mounting. This system helps in reducing them into convenient size which can easily be handle and stored like books and files. However, the problem relating to newspapers, in their storage and reinforcement stands still. The problem relating to newspapers is two-fold; first because of their large size and secondly, because of inherently poor material in making of news-print. It decays with faster speed, yellows and remarkable reduction in strength is recorded even in ordinary atmospheric conditions.

Though Indian conservationists tried to strengthen the decayed newspaper by coating an emulsion made of cellulose acetate and solvents. But the question is whether any such coating is capable to neutralize the higher acidity of news-print and stabilize them. The answer is negative. Then the question is whether or not any other suitable device for removing acidity and other impurities present in news-print is available so that they may be removed before coatings are applied. Devices are but they are very slow and do not ensure complete removal of unwanted imporities without damaging paper. One neutralizing technique which comes through applying alkaline solution of higher concentration is available. But it may suit to those whose collection of newspaper is small. Large stock of paper, however, difficult to treat with, smoothly, fastly and conveniently. In the backdrop of these facts, it can be said that any method of coating the surfaces of news-print for improving strength and stabilizing them are not lasting and wastage of time if it fails to neutralize acidity of paper. However, for preservation of this important source of information, the following system could be useful:

(i) Some copies of every news daily should be printed on quality paper and copy of each be supplied to National Archives of India, State Archives of the States/Union Territory where paper is being published and two should be retained for record by the Press. Some State Archives had shown interest on the scheme and sent proposals to their respective Governments and News Press which were principally agreed. But the subsequent development about scheme whether or not materialized, is not known.

(ii) The microfilming of newspaper is the system which gives effective answer to the problem. National Archives of India has started work on the scheme.

For preservation of Newspaper in a vast country like India, the Newspaper Archives could be an effective solution as it would lonely look into the interest of riches of newspaper published throughout the country.

(B) Problems of Individual Nature

One such problem of individual nature is relating to lamination of documents. Once the process, machine as well as solvent, was considered the key technique as it was the only technique to reinforce all types of records whether bitterly damaged or written with water soluble ink. But experience has brought up the facts into notice that lamination, specially solvent, carried out in the places reeling through humidity range 65% or above effects the legibility of the document. Second problem, however, is related to lamination of a document with low pH. Such documents, if laminated without removing acidic contents, create a problem in reversing the technique after few years. Thirdly, the laminated documents stored in humid and damped areas not only embrittle with fast rate but also cause source of indoor pollution emitting acetic and formic acid gases. Fourthly, the laminated documents stored in dry atmospheric condition brittle very soon. However, the problems of reversibility, emitting acetic gases and brittleness can be sorted out, if documents are laminated after neutralizing acidity and stored in monitored repositories.

Problems Relating to Restoration of Marathi Documents

Mr. A.N. Baji working with Marathi documents highlighted some problems in their storage and restoration. The first problem is relating to their size which obstructs in evolving any uniform method of their reinforcement. The paper or support of such documents is usually a long strip and width varies from about three inches to nine inches while no rule can be laid down for length. It may be anything from two inches to a hundred feet. They are long rolls which are made up by joining a number of long strips. The dimension of one typical roll, "A Mahzarnama" housing in Central Records Office, Hyderabad is approximately 8' x 120'. In contrast to such long rolls there are small strips of modern visiting card, each usually added as a postscript to some long letter.

The another peculiar feature relating to Marathi documents, is their writing which on the other side of a sheet begins, on that portion, on which the last line of the text on the obverse is written. To make it more clear, it is stated that the reader who has taken up the long strip to read, starts at the top and works down to the bottom. Then, to read the contents on the other side of the strip, he has to turn the page in the same fashion as one has to turn the page of a shorthand notebook.

The storage system of such long strips is unique. It is folded into convenient panels and its top edge is stuck on the recto of the inner side of the docket cover with help of a guarding strip. This system interferes with the free handling of documents. Further, Mr. Baji States that machine lamination of such long strips also pronounces many difficulties.

It is not clear whether any suitable system for the storage and reinforcement of such documents has yet been evolved. However, it is pointed out that in restoration of archival documents, we cannot aspect for any uniform system as individual document is an individual problem. So far as the storage, the documents in rolls can be put inside specially designed aluminium or board boxes fitted with pegs and pullies. It is convenient to store them in boxes. Supply of original rolls to users is not safer. The copies of the written contents on the obverse and reverse of the roll, should be got prepared in A4 size paper and bound arranging them in chronological order. Instead rolls, copies should be supplied to the readers. So far reinforcement, there is no problem in laminating the rolls in Emprignator laminator, the type being used in National Archives of India.

Problems in Preservation of Manuscripts in Twang Monastery

While working for conservation of manuscripts in Twang Monastery located in West Kumang District of Modern Arunachal Pradesh, few problems related to preservation and restoration of Kangyurs were faced.

Apart from moisture, a chief external agency causing deterioration of manuscripts by promoting growth of fungus and booklice, the rodents were of the serious concern. Traditional methods, such as trappers and baits, to get rid of the animal were ineffective because of the library were structure which could hardly obstruct their entry from surrounding. Besides, religious dogmas professing against the killing of lives in the monastery. Granary close to the library building and residence of monks and teachers located within the Monastery Campus, were functioning as the chief source of food for rodents and promoting their growth and existence. In such prevailing conditions, it was therefore, very difficult to annihilate them from the surrounding areas of library and to barricade their entry in the complex. As a measure to sort out the problem special steel racks, covered with non-rusting wire-mesh from three sides and flat door system from the front were suggested for storage of manuscript bundles in place of open wooden racks. This new system ensured free air circulation and security of the record materials from rodents.

Problem in Restoration of Multilayered Manuscripts (Tangyurs)

The unique structure of manuscript sheets of Tangyurs,[+1] getting place in Twang Monastery, posed a serious problem in their reinforcement. Such sheets were made of multithin paper layers. The thin layers of a Tangyur Manuscript sheet, stuck together by a local paste, had detached from various places. The reasons for such detachment may be several but main was the loss of bonding strength of adhesive. No doubt, the detached portions of the manuscripts sheet could have been corrected by repasting but in case of major work where permanence and durability was highly required, all the damages could not be corrected by mere minor mending. Also in case of bitterly damaged and decayed manuscripts, minor repair was not fit for giving lasting results. Even lamination (solvent) of such sheets was not possible because of air filled in the detached portions between layers and also because of the high humidity of the region. The captive air would have made a basis for delimination with the passage of time whereas high humidity would have turned the laminated sheet milky, adversely affecting legibility. Tissue repair was also not giving the right answer to the problem without correcting of detached portion. Moreover, thickness of sheet was also unsuitable for tissue mending. On the face of these problems, a new technique was devised for mending such sheets known as "special fulpasting". The process gave the right answer to the problem at the minimum cost. Details of reinforcement technique are available in Chapter 5 under special full-pasting.

Some Problems Relating to Miniatures

B.N. Tondon, in one of his articles, has brought to light some problems relating to reinforcement of manuscripts and miniature painting on paper. One of the problems related to the manuscripts is their mixed structure; written on one side and painted on the other. Such miniatures are from Tagore collections. The manuscripts are grappled with acidity. No suitable technique to deacidify them was found out because of the impure variety of colours used in general and the dye pronouncing pink colour in particular. The pink colour showed greater tendency to run away when wetted with water or alcohol inspite of the use of preservative.

Another problem concerning to the miniature paintings of Deccani and Western School was the gradual development of charring in the paper particularly at the back of green pigment. It grew to such extent that a portion of painting was only left with holes in place of beautiful green. Analysis of green colour had detected it copper acetate which is quite akin to the Mineral Verdigris. The actual mechanism involved in the process of charring could not be understood. However, it was suspected to be sulphurous/sulphuric acid. Deligibility of green colour in water and other solvents, was serious problem to neutralize acidity without harming the support and the painting. The solution of problem is yet to be found.

A dreadful problem which came to the light is related to red pigment made from Vermillion. This colour is gradually turning black. Its causes and means of cure are to be found out. Perhaps, the alternation from cinnabar to meta cinnabar, a black variety is responsible for this change. However, the proper cause as to how it has happened, whether due to some chemical action or otherwise, is yet to be understood.

B.N. Tondon, talking about a problem relating to miniatures states that a strange phenomenon was noted and that was the fast development of violet speck on white pigment. On analysis it was detected a lead continuing pigment. The technique of removal of speck or to convert it into original white is yet to be found out.

Flaking tendency in mass scale in a particular white pigment from zinc white, and cleavage and fading of illustrated manuscripts are some problems which need careful study and suitable measures to check them.

The problems relating to Bengal School of Folk Arts scroll painting, making use of numerous pigments and dyes are however, not easy. Out of the colours, yellow dyes which are affected by water as well as alcohol and some other organic solvents, spear head the problem. At the back of the paintings several layers of newspaper were fixed. These newspapers could only be removed by wetting in water but even slightest moisture will make the yellow dye spread quickly. Even several protective layers of polyvinyl acetate coating could not check the colours to run.

C.B. Gupta, in his article, "Conservation Problems in Tagore's Art on Paper" has mentioned about personal diary of Tagore which is fully illustrated. Some of the pages are illustrated on obverse and reverse while some on one side only. No unwritten margin was left on the paper sheets. Mr. Gupta says that the diary was bitterly damaged and each page has its own problem. Sheets are grappled with acidity.

As a course of treatment, to conserve them, it was decided to neutralize the acidic contents and subsequently mount them.

Ammonia and other non-aqueous methods were tried to remove acidity but they were not found satisfactory. The first reason was the nature of ammonia, deacidification which is temporary and not lasting. Second reason, however, was solubility of some colours in alcohol. As Mr. Gupta states, dry mounting of all the pages was not possible and the lamination pressing of individual painting was not desirable as the sealing does not leave any breathing space for paintings. As such, there was no alternative but to go in for deacidification by aqueous method and mounting with chiffon using zinc paste.

During the process of deacidification, using aqueous methods, the necessity of fixing red colour was felt. The other colours had been fixed with 1% PVA in Toluene. Number of solvents such as Acetone, Alcohol, Toluene, Benzen, Xylene and other solvents at different concentrations were tried but the bleeding of red colour could not be stopped. Besides, preservatives like Sandofix, Bedacryl, Methyl Methacrylate and Paralloid in aqueous and non-aqueous media in different concentrations were also tried. The Bedacryl and Paralloid in Benzene though checked the bleeding yet it left impression on the reverse of paper. Even PVA in trichloroethane and dichloroethane could not solve the problem of reverse impression though they completely stopped the bleeding. This problem still requires the help of conservators, to be sorted out amicably.

Problem Relating to Managing Effective Restoration in Archives

The effective conservation of documentary material can only be ensured if the best suitable treatment to counter the problem is given to the ailing document at the time when problem starts. Delay in timely treatment allows the document to suffer further. For managing the problem timely, the condition of all the holdings of a repository should be known to the conservation agency so that the priority of conservation treatment to be given to the ailing documents may be decided. Today, in most of the depositories, practice is that those affected documents, which either incidentally at the time of tying bundles or attending search cases or otherwise come into notice, are sent to workshop for their treatment. Author has experience that sometimes damaged files are sent to laboratory for correction only after it is pointed out by the users. It is done perhaps more because of research rule provisions governing supply of records to users and not with the passion that ill-records should not be supplied to users. Supply of damaged records to the users may add fuel to the fire. In other words, the supply of records, poor in health, to the users, will bring about more deterioration to them. Moreover, there is no any definite working system for access to the ill-records in a repository. As a result, timely treatment to such records is denied. It is one of the greatest problems because of which conservation of documentary heritage is suffering inspite of availble suitable aids.

The effective solution to the problem is to classify all the records of the repository in following three categories; either agency wise or stack area wise.

(i) Bitterly decayed and damaged records seeking urgent chemical treatment and repair. If possible, the nature of damage or infestation with what document/file is suffering should also be indicated.

(ii) Records in poor state of preservation.

(iii) Records showing the sign of fast decay and damage because of acidity.

While classifying the records the information about stack Area No., Agency No., Rack No., Bundle No. and File No. should clearly be indicated in the list helping access to records without any loss of time.

REFERENCES

Baji, A.N., "Some Problems Relating to the Preservation of Marathi Documents", *The Indian Archives*, Vol. XII, Nos. 1-2, 1958, p. 76.

Gupta, C.B., "General Conservation Problems in Tagore's Art on Paper", *Conservation of Cultural Property in India*, Vol. X, 1977, p. 54.

Gupta, Motilal, "Difficulties found in Collecting and Conserving Cultural Materials, Manuscripts", *Conservation of Cultural Property in India*, Vol. XII, 1979, p. 54.

"Indian Historical Records Commission, A Restrospect", Vol. II, 1948–1994, p. 10.

IS: 220-1988 Ferrogallotannate fountain Pen inks 0.1% iron content.

IS: 1581-1975 Ferrogallotannate fountain Pen inks 0.2% iron content.

IS: 5805-1993 Specifications for ball point pen inks.

IS: 1551-1959 Carbon Paper for typewriters.

IS: 4174-1967 Typewriter ribbon ink.

Perti, R.K., "Some Problems in Conservation of Documentary Heritage in Indian Repositories", *Conservation of Traditional Records, Paper and Allied Materials*, p. 37.

Prajapati, C.L., "Conservation of Tawang Monastery", *The Indian Archives*, Vol. XXXIX, No. 1, 1990, p. 23.

Prajapati, C.L., "Modern Techniques on Conservation of Palm Leaf Manuscripts", *Proceedings of Palm leaf and other Manuscripts in India Languages*, p. 201.

Prajapati, C.L., "Problems Relating to Establishment of Effective Preservation and Restoration in Archives and Libraries", Lucknow, Librarian, Vol. 27, Nos. 1-4, 1995, p. 37.

Proceedings of the 10th Meeting of National Committee of Archives, Trivendrum, 30th December 1958, p. 41.

Proceedings of the Fifth Meeting of National Committee of Archives, Simla, 6-7, July 1956, p. 3.

Tondon, B.N., "Problems in Conservation of Miniature Paper Paintings and Illustrated Documents", *Conservation of Cultural Property in India*, Vol. X, 1977, p. 38.

+1 Tangayurs are the translated texts of the Commentarial works as well as other secular treaties written by Indian Pandits and other non-Tibeton scholars. Most of the sheets of Tangyur Manuscripts are multilayered or each sheet is made of two or more than two thin sheets stuck together with suitable adhesive.

7

ROLE OF INDIAN HISTORICAL RECORDS COMMISSION FOR THE CAUSE OF CONSERVATION OF RECORDS

Indian Historical Records Commission, which came into existence in 1919 for the cause of Archives, had played a key role in the development of modern conservation in India. So far its objectives, in the words of R.C. Majumdar, the Records Commission was constituted for the purpose of advising archives on the best means of preserving its records, listing, calendaring and making them accessible to the public.

Tracing growth and development of Commission, it can be said that the second report of Royal Commission on Public Records and a Memorandum of Mr. William Foster of the India Office (1914) seem conceiving its seeds. The report as well as memorandum, reiterated the need for reviewing the prevailing condition of central and provincial records and pursuing for an uniform policy dealing with them. Meanwhile, the Government of India was lacking a Committee of Experts to help custodians in dealing with the subjects professionally technical.

As time went by, Raibahadur Jaimini Mohan Mitra, then Assistant Secretary in the Department of Education, made a half hearted suggestion to constitute a small committee of archivists and historians to deal with the problems related to records at ordinary times. But 1917, because of war, was not admitted normal times. As such later on, Sir Edward Maclagan, Secretary to the Government of India, in the Department of Education, referred the question to Professor Ramsay Muir who strongly recommended the appointment of a Historical Material Commission with its headquarters in Delhi. He also recommended that the Commission should have provision to induct the Office-in-Charges of records in each of the British Provinces as well as in Principal Native States from India and the Chief Executive Officer, a trained historian and archivist from Europe. The idea got favour with Government of India and blessings of the Secretary of State in due course of time and finally implemented by Resolution No. 77 (General) passed on 21st March 1919. Thus, the Indian Historical Records Commission took birth which in subsequent period started to roll on.

The Commission since its birth to date has seen various ups and downs. It was reorganized and reconstituted from time to time. But after 1948, the charter and composition of the Commission underwent a sea change with an objective to fulfil the demand and aspirations of free India. The need of the Commission to have a permanent Committee to monitor and implement the Resolutions passed by the Commission and also to advise during the intercession period was being felt from a long time and that was fulfilled in March 1972 by reconstituting the Commission. In the years to come, after a period of 18 years, once again the Commission was reconstituted in 1990 which has made provision to appoint the Union Minister of Human Resource Development as the head of the Commission.

Role of Commission in Promoting Cause of Conservation

During its period of 80 years (upto 1999) the Commission, functioning as apex body in the field of Archives, took-up landmark decisions and passed resolutions in many areas of archives conservation, in both pre-independence and Post-Independence period. All such resolutions were aimed to save the priceless treasure of cultural heritage in India and to make it available to the scholarly world.

The precarious condition of district records and the

incharge of ill paid ministerial officers absolutely innocent of the elementary principles of archives keeping (a part of archives conservation) was known to the Commission. The storage conditions were seldom satisfactory. The design of buildings housing records was unsuitable from the stand-point of preservation and lacking desired equipment. As such, a resolution was passed, at a very first session of the newly reorganized Commission (1942) to draw attention of Provincial Governments towards urgent need for concentrating district records. The Resolution addressed to the Government of Bengal, Bihar and Orissa, in particular and all Provincial Governments in general (except Madras and Bombay) focused attention to the ill condition of records which in many cases had been reduced to pulp by ravages of insects and the deleterious elements in the climate, causing permanent damage to the paper and ink.

It is universally true that records once bitterly damaged or deteriorated are lost for ever. Therefore, prevention is always preferred than cure. This is the reason that preventive measures are preferred over those of rehabilitation. If records are well preserved, the rehabilitation can be resorted ahead to centuries. Keeping this fact in view, the Commission repeatedly urged upon the provinces to the need for organizing central record offices providing optimum conditions for storage of their old papers. So for as the implementation part of resolutions passed by the commission it was slow, almost in all the provinces, as the country was not aware about the importance of records. In other words, the people were lacking archival sense. Yet, whatever progress in maintenance of records in central record offices was made that goes entirely into the credit of Commissions endeavours.

Commission's attempts in salvaging manuscripts in private custody, in the meanwhile, were wonderous. In a country with India's hoary antiquity, a hung bulk of historical manuscripts was bound to lie unnoticed and uncured in private custody. The owners in many cases had neither the requisite means nor the required knowledge for their storage and preservation. But in such cases individual delinquency whether international or otherwise was bound to bring about a national loss. The Commission, therefore, decided that the historical manuscripts in private ownership must be brought to light and countrywide survey must be organised with the objective of salvage.

During the days of crucial war, Central and provincial Governments were too preoccupied in military preparations to spare money and manpower for such survey, yet the task brook no delay. Tons of old manuscripts irrespective of their literary or historical value in order to meet the scarcity of paper were being repulped. In other belligerent countries, private sector had been enlisted to prevent any type of literary loss but in India Commission was only organized body to take care of such happenings.

So far as rendering advice in the field of preservation and records keeping, the ex-officio Secretary of the Commission sought that Keeper of Records (Now Director General), Government of India, should visit custodial agencies from time to time for the purpose. In 1940, the Keeper of Records, on invitation from the Government of Bombay visited and inspected records housed in the Alienation Office at Poona and advised on the best methods of preservation of cultural heritage known at that time. Commission also sought that Keeper of Records should visit once each province in a block of three years and should submit report to the Government of India and the Historical Records Commission as well on the state of preservation of manuscripts and the measures to be taken up for their improvement. Besides, suitable steps should also be taken up for disseminating technical know-how to the organizations dealing with old records and manuscripts.

The re-constituted Commission in its very first Session at Mysore in January 1942 urged the necessity of bringing out a quarterly or biannual archives journal to make people available the knowledge of the science dealing with preservation of cultural heritage. It was an unique attempt by the Commission to disseminate technical know-how to the institutions dealing with documentary heritage and to the people interested on the subject.

The Post War Reconstruction Scheme, however, is marked for its significant outstanding decisions in the field of documentary conservation. The first decision taken in this regard was to develop Preservation Section of National Archives of India, in the lines of the record offices of world in advanced countries. The second decision was concerned to prohibit unwarranted destruction and export of historical manuscripts. The third such decision taken by Commission was related to arouse awareness about conservation among the officials of record rooms in British India and native provinces. The most significant development taken place in pre-independence period was to start experimental studies in the field of preventive conservation. The glaring example in this respect is related to take up the study of climatic condition in New Delhi from the stand-point of preservation. Taking up the work, the relative humidity and temperature of stack areas at ground floor and second floor of National Archives of India were recorded from 1940 to 1943. On the basis of variation, flow chart of both the elements, it was concluded that the climate of Delhi is not at all suitable for old records and manuscripts. Viewing over the fact, the Commission came out with the view to go in for air conditioning of the muniment rooms. This facility would not only provide uniform temperature and humidity at optimum range but also keep the insects under control partly if not fully. However, to thrash the repairing work of records, the extermination of pests from them is a must. As such, for the purpose, fumigation of records, old and new, was emphasized. To restore decayed and fragile records, the need of a laminating machine was felt. In order to meet these requirements, Commission recommended the installation of vacuum fumigation and laminating machine in the Imperial Record Department at an early date.

The Commission in pre-independence period, whereas aroused awareness among the senior staff of Archival organizations, provided a suitable infrastructure for healthy growth and quick development of Conservation Science in post-independence period. It can be visualized from the glimpses into the major achievements of the Commission in the various areas of conservation in post independence period.

Functional Archival Building

The Commission had realized the importance of suitable

functional buildings for proper preservation of records. Few State Archives, though had organized archives, the condition of records was appalling. Buildings, wherever they exist, were ill-designed and ill-equipped. National Archives of India had a better building than archives in provinces, but paucity of space shortage was being felt since 1945. Therefore, there was a pressing demand for an annexe building. The matter was discussed at 23rd (1946) and 24th (1948) sessions of the Commission but until 1959, the progress remained tardy. However, the Commission continued to voice its concern and expressing its anxiety re-emphasized at its 37th session, to give top priority for construction of functional archival building. As a result of persistent efforts of the Commission, the funds were made available for the construction of functional building. In 3rd Five Year Plan foundation stone for construction of annexe building was laid in 1979 by the then Education Minister at the Centre. Subsequently, in 1982, administrative approval for its construction was received. The first phase comprising of seven storied Stack/Tower Block was completed during the period of 7th Five Year Plan and dedicated to the nation by the Union Minister of Human Resource Development and the President of the Commission on 3rd December 1991.

It was due to the sustained efforts of the Commission that by 1975, the State Archives of Andhra Pradesh, Orissa, Rajasthan, Tamil Nadu and Uttar Pradesh had their own functional buildings. Some other states like Assam, Bihar, Delhi, Goa, Gujarat and Kerala have also since joined this group. The matter is being pursuaded by other States vigorously with their respective Governments.

Preservation of Important Records Created by Agencies other than Central and State Governments

The Commission's role in preserving records created by agencies other than Central and State Governments may be considered a well timely taken step which prevented the loss of valuable records lying with the custodians of society. The Archives at Centre and States were mostly confined to preserve the records created by Central and State creating agencies while the records created by former rulers, *jagirdars*, officials and various municipal and local bodies lying in the custody of temples, maths and mosques were hardly getting due attention of archival agencies.

The Commission, well aware of the fact, stressed upon the need for their proper maintenance and preservation from time to time. In its 41st session held at Trivandrum in 1971, Commission passed a resolution (No. VIII) urging authorities to maintain records in custody of religious places like temples, maths, mosques and churches, etc. and to ensure their proper preservation, Commission also urged from religious authorities to deposit records of their custody in Archives or learned institutions for preservation. Subsequently, in its sessions at Panaji, in 1973, Commission passed a Resolution (No. XIII) in which Municipal Corporation and other local bodies were urged to preserve their archives in properly organized record rooms. In 1979, Commission *vide* Resolution-IV, passed at the 46th Session held in Aurangabad, urged State archives to preserve records of former rulers, *thikanedars, jagirdars* and officials.

The recommendations of Commission, through resolutions, now started to bear fruits. In pursuance of Resolution-VIII of the 41st Session, the State/Union Territories of Delhi, Goa, Gujarat, Karnataka, Nagaland, Orissa, Punjab, Rajasthan and West Bengal appealed to the authorities of temples to transfer their records in the Archives Departments for proper preservation. The proposals sent by the State Archives of Andhra, Haryana, Kerala, Madhya Pradesh and Tamil Nadu were in active consideration by their respective Governments.

Haryana accepted the directives of Resolution XIII of the 42nd Session, in principle, while the matter was being considered in Andhra Pradesh, Delhi, Gujarat, Karnataka, Madhya Pradesh, Maharashtra, Manipur, Punjab, Tamil Nadu, West Bengal. Goa, and Himachal Pradesh, have acquired records of local bodies, while Rajasthan state Archives had taken-up the work of inspection of records of these organizations.

As envisaged in Resolution-IV of the 46th Session, the records of former rulers, *thikanedars, jagirdars*, officials, etc. have been acquired by the State Archives of Andhra Pradesh,

Karnataka, Kerala, Maharashtra, Orissa and Utter Pradesh. The Government of Tamil Nadu had ordered to respective organizations to transfer the records of former rulers (upto 1857) to the State Archives while that of later period to maintain in respective Departments. The Resolution has been accepted by the Government of Rajasthan, while the same is under consideration in Goa, Madhya Pradesh and West Bengal.

Preservation of Cinematography Films

Cinematography films, throw light on the tastes and values of the society of their contemporary period and hence are considered the important source material for socio-cultural studies. In Indian sub-continent, this valuable wealth had attended focus least for their proper preservation while in Western World, it was being given due care. It can be visualized from the fact that Film Archives under the aegies of their respective National Archives had been setup. India diverted its attention to the task first time in 1982 when a resolution at 48th Session was passed by Commission, to place the task of preservation of Cinematography films under the aegis of the National Archives of India. But due to unwillingness of Ministry of Information and Broadcasting, the proposal of the Commission could not be executed. The Commission now felt that the Film Archives could continue to function under the Ministry of Information and Broadcasting but efforts should be made to bring National Archives of India and Ministry of Information and Broadcasting closer for preservation of this wealth. Working on the theme, Standing Committee at its 11th Meeting, placed before the Commission at its 49th Sessions (1985) that the Director (Now Director General) of Archives could be associated with the Committee of Ministry of Information and Broadcasting, for rendering professional and technical guidance to Film Archives in connection with preservation. The Commission endorsed the view of Standing Committee and Ministry of Information and Broadcasting was approached but no tangible results had emerged.

Secretary of the Commission, however, in older to find out an amicable solution discussed the matter with the Director, National Film Archives, Pune and he found that the Director, Film Archives, was keen to have meeting with the Director,

National Archives of India. Secretary of Commission while apprising members of the discussion of 15th meeting of Standing Committee held on 3rd March 1992, expressed his hope that the question of associating Director General, National Archives of India on the Advisory Committee of the Film Archives would be resolved soon.

The number of discussions which took place with Ministry of Information and Broadcasting about the matter of placing National Film Archives under aegis of National Archives of India for Preservation of Cinematography Films, resulted in arousing awareness among the authorities of Film Archives, Pune and they started to take interest on the subject. Now Film Archives, Pune is functioning well. They are the first claimant to have Compactus shelving system in India. A Preservation Officer, is looking after the work of preservation of films. However, the task of conservation of posters in their custody is still demanding active consideration, failing which the rich culture depicted through them may go to oblivion.

Development of Quality Paper and Ink for Creation of Permanent Records

The need of Standard quality paper and ink, when was first time realized in India for creating records of enduring value, is yet to know. But for the first time as early as 1951, the Local Records Sub-Committee of the Commission at its 13th meeting urged the Director of Archives to submit proposal for different kinds of paper and ink that could be used by the Government Offices for creation of records. In the following years, National Archives of India submitted a proposal suggesting various grades of paper and ink suitable for creation of records of permanent value. The proposal was forwarded to the Controller of Printing and Stationery and Pulp and Paper Section of the Development Wing of Ministry of Commerce and Industry but no appreciable success was achieved in getting writing material of quality. The Commission, however, continued to voice its concern and passed resolutions at its 31st Session (Resolution IV 1955), 36th Session (Resolution III 1961) and 42nd Session (Resolution XII) to expedite the matter.

The Commission, subsequently resolving in its 43rd Session at Lucknow, passed a regrettive note. In pursuance of the constant demand of the Commission, National Archives pursued the case with the Government of India which constituted a Committee in 1975 under the chairmanship of Additional Secretary (Culture) to consider the case. The Committee in its meeting in May 1975, wherein the representatives of the paper manufacturers were also present, got success in pursuing paper manufacturers to supply samples of good quality paper to National Archives of India for testing their suitability.

Papers received from Ballarpur Paper and Strawboard Mills; J.K. Mills, Titagarh Paper Mills, Bengal Tiger Mills were tested in the Conservation Research Laboratory of the Department and on the basis of the test results, paper manufacturers were suggested to bring out improvement in the papers. In 1978, the 'Sun-lit-Twist Wrap Bond Paper', an improved quality paper manufactured by M/s Ballarpur Paper Industry was supplied to this Department. The paper was tested and found very close to I.S.I. specifications laid down in IS: 1774-1986, "Specifications for Paper for Permanent and Semi-permanent Records". Unfortunately, the attempt of National Archives of India to procure this paper from the Controller of Stationary, Kolkata and Director General of Supplies and Disposal, New Delhi, proved in vain. National Archives of India, however, in its fresh move, requested paper manufacturers like M/s Triveni Tissue Paper Ltd., M/s Sehgal Paper Ltd., M/s Ballarpur Paper & Strawboard Ltd. and Hindustan Corporation to manufacture paper of I.S.I. specifications. Meanwhile, Commission at its 52nd Session in 1988, at Srinagar re-raising the issue, passed a resolution urging early action on the matter. Keeping in view the desire of Commission, National Archives of India, reminded the afore-listed paper manufacturers to supply the paper of quality for testing. The attempt this time remained successful and samples of paper were received and tested in the Conservation Research Laboratory of the Department. The papers this time produced by Ballarpur Paper and Strawboard Ltd. and M/s Triveni Tissue Ltd., were suitable for creating records of A and B category.

Standard specifications, relating to fountain pen ink, to be used for creating permanent records are available under

IS-220-1988, IS: 221-1962, and IS-1581-1975. In a successful attempt, National Archives of India in collaboration with the Bureau of Indian standards, has also finalized specifications for Ball-point pens and Ball-point ink (IS-5805-1993) for use in creating agencies. But, the question, how paper, fountain pen ink and ball-point pen ink of standard specifications could be made available to users is yet to be solved.

Scheme of Financial Assistance for Preservation of Records in the States/Union Territories and in Private Custody

The shortage of funds for running conservation activities is an universal phenomenon. Recently, when Dr. Christine Makey of Wales Museum, visiting India and delivering a lecture in the workshop organized by National Archives of India in January 1999 was asked about the greatest problem faced by her in the conservation activities, she came out with reply that of "funds". It can be presumed, when countries like United Kingdom are over-shadowed by the problem of funds, what would be the position of developing countries like India. Briefly speaking, the money is the major bull locking horn in the activities carrying out conservation in Archives and allied institutions in India. It is not appropriate to discuss the reasons responsible at this juncture. However, it is noteworthy to mention if a horse carrying load suiting its capacity is sacked by additional load, the speed of the horse would definitely be retarded if back-bone does not rupture. The situation of Archives in India had been very much akin to the problem relating to the horse. The financial problem of Archives, becomes much more tightened when they are assigned the job of conserving manuscripts of religions and other natures lying in private custody, on their own budgetary provisions. To sort out this problem, Commission passed resolution No. III in 52nd Session held at Srinagar in 1988 for financial assistance to custodial agencies commencing from 8th Five Year Plan.

In pursuance of this resolution, a proposal was forwarded by National Archieves of India to the Department of Culture. Government of India, for formulation of a new Scheme for financial assistance to facilitate conservation activities of records in private custody located by the Regional Records Survey Committee and to have separate budget head for this purpose.

Department of Culture approved the Scheme and also expansion of Scheme of financial assistance for the Department of State/Union Territory, Archives under provision for "Repair and Preservation of Manuscripts/Records" in private custody, listed by the Regional Records Survey Committee.

Preservation of Vital Records

Records relating to birth and death are considered an important source material for demographic studies world over. In India, records of this category created by the Registrar General and Census Commissioner of India were not getting due consideration from the view point of conservation. Even a very little was known to the scholars about their availability. Consequently, to provide conservation umbrella to such records, the 48th Session of the Commission held at Gandhinagar (1982) adopted a recommendation suggesting that vital records pertaining to births and deaths be properly preserved and their microfilms be made available for research after blackening out the cause of death. Commission, further recommended that the case of format for microfilming of such records could be discussed by Standing Committee with the Registrar General of India and be finalized by the Committee of the Commission. The matter was accordingly taken up with the Registrar General of India who intimated that the registers of birth and death are the records of state and Municipalities and are not covered under the Archival Policy Resolution. However, he welcomed the idea of microfilming of such records.

The whole matter was then placed before the Standing Committee of the Commission at its 10th meeting, where it advised to devise the format for microfilming of such records in consultation with Registrar General after compiling data completely. The matter was discussed with the Registrar General who intimated that the work of compiling data is not yet complete. Further, he expressed his inability in undertaking the work of microfilming because of financial constraints. Subsequently, the matter was placed before the Standing Committee at its 15th meeting. The Chairman of Standing Committee viewing over the importance of records suggested to take up the issue with the Registrar General after a legislation on Public Records was enacted. The matter is actively being discussed as the enactment of law has been taken place in 1997, but nothing tangible is yielded so far.

The glimpses of activities covered by the I.H.R.C. (Indian Historical Records Commission), since its birth to date in the field of Conservation of Records, are self-explanatory. It shows that the Commission had been actively involved in arousing awareness about the importance of records, necessity of their conservation, and disseminating the knowledge to the custodial agencies. Not only this, it also acted as a directive authority in India forcing other custodial agencies to take up the task of conservation through National Archives. Perhaps, the lack of such an apex Body, would have retarded the fast growth and development of conservation in the sub-continent which in turn would have resulted in loss of valuable information for ever.

REFERENCES

Indian Historical Records Commission, A-Retrospect (1919-1948), Vol. 1, p. 3.

Indian Historical Record Commission Proceeding, Vol. 28, 1951.

Indian Historical Record Commission Proceeding, Vol. 36, 1961.

Indian Historical Record Commission Proceeding, Vol. 40, 1970.

Indian Historical Record Commission Proceeding, Vol. 41, 1971.

Indian Historical Record Commission Proceeding, Vol. 42, 1973.

Indian Historical Record Commission Proceeding, Vol. 43, 1975.

Indian Historical Record Commission Proceeding, Vol. 46, 1979.

Indian Historical Record Commission Proceeding, Vol. 49, 1985.

Indian Historical Record Commission Proceeding, Vol. 50, 1987.

Indian Historical Record Commission Proceeding, Vol. 52, 1988.

Indian Historical Record Commission Proceeding, Vol. 53, 1990.

"Post-War Reorganizations of Archives Office, in India", *I.H.R.C. Proceeding*", Vol. 21 Dec. 1944.

"Status Report on Resolutions Passed" by the *Indian Historical Records Commission* (40th-52nd Session)-Rohtak-20-21 Apr. 1992.

8

CONSERVATION OF *VAISHNOVITE* MANUSCRIPTS ON *SANCHIPAT*

Majuli, resting on the lap of river Brahmaputra, is known to be the largest river island in the world. The island, marked for its natural beauty and riches of cultural treasure depicted through its various satras in the form of Vaishnovite Manuscripts and Museum objects of rarest nature, geographically falls in the Jorhat district of modern Assam. The area of island which was once 1250 sq. km. is now reduced to 750 sq. km. due to soil erosion.

The satras are the Vaishnovite Monasteries for religious practices as well as the Centre for Art and Culture. The objective of satras is to motivate the mids of the people to realize the existence of supreme power, God, through the ritual performances and through performances of prayer, *dharana,* dancing, *nam prasanga* and recitations of spiritual power of lord Krishna.

The religious theme emerging out of satras, in a way is interesting to bring about cultural unity of the people of Assam and that can be visualized by the number of satras, about 600, spread throughout the state. In Majuli itself, 65 Satras were active at a time but now the number is reduced to 22 due to continued erosion and flood of river. Some satras have been shifted to other parts of Assam.

Out of 16 Satras, Author visited 9 satras while remaining could not as the access to few satras was not possible because of communication problem.

Table 8.1 Satras of Majuli

Satras which were Visited	*Satras which were not Visited*
1. Dakshin Path Satra	1. Narasingh Satra
2. Bhogpur Satra	2. Punia Satra
3. Garmur Satra	3. Belo Guri Satra
4. Baingna Ati Satra	4. Bihimpur Satra
5. Auniati Satra	5. Dighali Satra
6. Natun Kamlabari Satra	6. Natun Chamaguri Satra
7. Shri-Shri Uttar Kamala Bari Satra	7. Sri Anirudh Dev Krishti Sangha
8. Bolowsudhia Satra	
9. Natun Chamagarhi Satra	

Observations

The information carries such as palm leaf, Birch Bark, papyrus, Vellum, Parchment and paper are well known to many but few know about the Sanchipat and Tulipat abundantly used by Indian scholars for the purpose of writing in the North-east. Though, Author had learnt about these from the fragments of Cultural sources and had seen few manuscripts housing in State Museum, Manipur, yet he was amazed and spell bound observing the riches of manuscripts on the carrier housing in satras of Majuli. The size, quality and riches of collections are really a wonder and matter of pride to India as she is only the claimant of the material deployed for the purpose of preserving knowledge even in modern times.

The smallest size of Sanchipat manuscripts measures 12.5 x 3.25 cm. while largest 65 x 27 cm.

The ink deployed for writing of the manuscripts is black. It is difficult to mention as to how many varieties of inks were used through the period for writing over manuscripts, but according to Shri Kanaka Datta ex M.L.A. of Majuli informed, a local ink prepared by the people of Satras was used for the purpose. The ingredients of the ink referred to above were,

Haritri, Ghamira and cow urine. As my knowledge goes, it is the ink so far not known to the people.

Sanchipat is the bark of Agru tree, popularly known to the people of Assam as Sanchi Tree (in Assamese language). The bark of the tree as it cannot be deployed for recording or depiction and therefore, subjected for series of mechanical and chemical treatments to make it liable for writing. The subject matter of writing of most of the manuscripts on Sanchipat is religion (Krishna Charita), astronomy and medicine.

Manuscripts housed in Dakshin Path Satra (Established in 1662 A.D) are charred and turned dark brown from edges. Corners of some manuscripts are missing while some are damaged. Some of the manuscripts wrapped in cloth are tied between wooden boxes. The bundles of manuscripts are kept one over the other unsystematically, gathering dust and insect beat. (See Figure 8.1). Briefly speaking, storage system is squirely inappropriate and manuscripts are in poor state of preservation. It seems that manuscripts so housed have neither been used nor been cleaned off from many years.

Various manuscripts are in the custody of Bhogpur Satra. (Established in 1528 A.D.) as well as in the personal custody of Bhagats of this Satra. Gunamala (Summary of Bhagwat Geeta) Ankya Matya, Ratnavali and Madhya Darshan are few manuscripts, which I examined thoroughly. The sheets of Gunamala Mss. made of Tulipat[+1] are turned yellow, and damaged from edges and corners. Sheets of other manuscripts on Sanchipat are damaged by silverfish, cockroach and booklice. Few upper sheets of manuscripts are badly stained by cockroach beat and dust. Some sheets are infected by a variety of fungus. Interior sheets of most of manuscripts kept in wooden boxes are abit moistened.

Most of the manuscripts of Garmur Satra are disarranged, badly damaged, charred and eaten by silverfish. Some are infected by fungus while some, kept in wooden boxes, are moistened. Few manuscripts are infected by a variety of black fungus. According to Mr. Kanak Datta, ex-M.L.A., now Secretary of the Committee constituted to look after the cultural artifacts including manuscripts of the Satra; a manuscript "Dashamkhand Bhagwat", is older than 400 years. The unique thing came into notice during examination is the reinforcement of few damaged sheets of manuscripts using a very thin sheet of Sanchipat. Though, restoration work attended is a minor patch work, seems to be of quality in all respects.

The manuscripts of Bangana Atri Satra are also on Sanchipat but one manuscript is on Tulipat. Comparatively to all other satras, manuscripts of this satra are well arranged and well kept in steel almirahs. Storage system is really attractive and impressive. See Fig. 8.2 & 8.3. Manuscripts however, are dark brown from edges while corners of a few are missing. A Manuscript of illustrated *Mahabharat* on thick Sanchipat is in good condition. Few manuscripts require dusting and repair.

Auniati Satra (Established in 1653 A.D.) is noted for rich collection of manuscripts of Sanchipat. Written margin of illustrated *Dashamkand Bhagawat*, is charred and yellowed from the side margins. Sheets of few manuscripts kept in wooden boxes are abit moistened.

Most of the manuscript sheets housing in Natun Kamla Bari Satra (Established in 1936) are infected by fungus and charred. The oldest manuscripts according to Shri Rammnath Barua, a bhagat of satra, is *Shrimat Bhagwat*, written by Shrimant Shankardev and bears a date on its back more than 400 years. Approximately 350 manuscripts, including manuscripts in personal custody of Bhagats of satra are available in pothikhana. Some of them are kept in wooden boxes while some are tied between two wooden boards.

About 60 manuscripts are available in Pothi Bharal (Library) of Shri Uttar Kamla Bari satra which was established in 1673 A.D. Sheets of manuscripts are charred, fungus infected, insect eaten and moistened.

Manuscripts of Balasudhia Satra (Established in 1675 A.D.) are mostly damaged by water and lying uncared. Because of fungus infection sheets of few manuscripts are stuck together and have become hard losing flexibility. The manuscripts can be restored if work is attended by skilled restorers in the supervision of experienced expert.

Summary of Observation: Summing up the whole episode of observations made above, it can be said that the manuscripts in the custody of various satras of Majuli are suffering adversely because of dust, fungus, insect and moisture. The decay and damage is aggravated by poor storage, water (flood) and lack of knowledge of preservation. The storage system is unscientific. Most of the satras are using wood boxes for their storage. Some manuscripts have been put between two hard boards, wrapped in cloth and then stored in boxes. One satra, however, have made efforts to arrange manuscripts systematically and store in steel almirahs.

Recommendations for Proper Preservation of Vaishnovite Manuscripts

For the preservation of Vaishnovite manuscripts of the satras, following measures are suggested on the minimum requirement basis.

(i) Dust and beat of insects accumulated on manuscripts should be cleaned off using cotton wool or cotton cloth or any other convenient device such as brush etc. Subsequently, they should be wrapped in cloth of either colour, red or yellow and tied between wooden slabs larger than the size of manuscripts or kept in wooden boxes.

(ii) Wooden boxes being used for storage of manuscripts, wooden boards and cloth for wrapping them should be dried in sun before onset of rainy season and after it subsides. Moistened manuscripts, however, should be exposed to open environment and dried in shadow. Subsequently, these should be arranged in their original order, wrapped, tied and stored in wooden boxes as suggested above.

(iii) To discourage insects, the margosa leaves (*Neem* leaves) dried in shadow should be placed on the bottom of the wooden box, interleaved with manuscripts sheets and on the top of the manuscripts bundles and wrapped with cloth. The leaves in question should be replaced by fresh lot after every six months or so. Besides, if available, commercial insect repellent, Naphthalene pieces may also be kept in boxes or wrapped in cloth with manuscripts sheets to get rid of insects.

(iv) Quality silica gel, after tying the crystals in a piece of cloth, should be kept in the boxes along with manuscripts sheets for absorbing excess of moisture within the box. It should be dried in sun after each rainy season and kept again.

(v) Bundles of manuscripts, preferably be kept in steel almirahs. If steel almirahs are used for storage, Naphthalene bricks and silica gel should be kept in shelves.

(vi) Water soaked manuscripts sheets should be dried in shadow arranged chronologically wrapped in cloth and tied in wooden planks or kept in wooden boxes. These should not be left loose exposed to environment for a long period of time.

(vii) Fungus stains, could be removed using 10% solution of thymol in methylated spirit. As preventive measure, thymolised paper should be kept on the top and bottom of the manuscripts bundles.

(viii) Bitterly damaged sheets should be restored using Sanchipat and suitable paste.

Investigational Work to be Attended

Keeping in view the conservation of this wealth, it has been decided to take investigational work on the ink and support material Sanchipat as under:

(i) Out of the ingredients of ink indicated on p. 268, effort is being made to prepare the sample ink and to study its adverse effect, if any, on the plant fibre of Sanchipat and paper.

(ii) Besides, study on the effect of chemicals used for removing fungus stains, separation of blocked sheets and suitability of repairing techniques being used in Archives at present for reinforcement of water soaked sanchipat manuscripts will also be taken up.

REFERENCE

+1 A writing support, used in large scale in Assam. It is generally found in three colours, red, grey and white. For preparing grey coloured Tulipat, the bark of *Jamun* tree is used while white is prepared from the bark of *Mabai* tree. Barks of respective woods are treated suitably to make them liable for writing. A written sheet on Tulipat is known as "tulipateeya."

9

ANALYSIS OF WORK DONE IN THE PAST AND NEED OF PERSPECTIVE POLICY CHANGE

The present study in-depth on various aspects of conservation, reveals the fact that conservation theory and praxis are deeply rooted in Indian sub-continent and have some philosophical, scientific and cultural trend. This unique combination is the rare feature of the Indian conservation today and is consonant with its traditional glory. Briefly speaking, it provides suitable know-how for conservation of documentary heritage and is capable of meeting any problem relating to it. However, the picture which emerges out of the other side of the coin, is that a good chunk of riches of documentary heritage, scattered throughout the country, in public and private custody, is insecure from onslaught caused by various factors. It is a matter of concern for all of us.

The riches of documentary heritage not only in the remote areas of the country but also in the areas where computers, the modern magic eye work wonders, are dying unattended. They grapple with dust and insect, await careful attention of the conservationists. This neglected heritage never commands respect from even elite visitors because of its unsystematic arrangement and unscientific storage. Some of the neglected documents have already reached to the brink of disaster while some are just to reach if due care is denied for few years more.

There are many examples of neglect, but one recently got reported in the *Indian Express*, 12 November 1997. The paper reports:

"The historical treasure of Jammu & Kashmir State, housed in the Lal Mandi State Museum, Srinagar, is decaying as the Archives Department of the State has dumped thousands of antique which include material on art, geography, armory, paintings, textiles, manuscripts, etc. Further, it tells that the primitive methods of preservation employed by the State Archives are not capable to sustain manuscripts for long. The copies of rare manuscripts are available in the British Museum and nowhere else. Apart from age old method of preservation, the next problem is the non-availability of qualified staff."

It is unexpected when the people of custodial agencies have developed some archival sense and are aware of the importance of documentary heritage. Skilled restorers, conservators and conservation scientists are infact available today. It is not expressed hypothetically but on the basis of experience and facts acquired during my days in school of Archival Studies as teacher of Conservation. The School is training at least 30 to 40 candidates in the field, sponsored and non-sponsored, at professional and sub-professional level, every year. No doubt the percentage of trainees, specially private, from far flung areas of the country is lower. Moreover, financial assistance to the custodial departments for conservation of cultural heritage by National Archives of India and also by the Department of Culture under Ministry of Tourism and Culture is provided. Besides, advisory services are being extended by the department to needy organizations. Awareness about archives, archival sense and conservation of the wealth is created by organizing workshops, seminars, exhibitions, and delivering lectures and talks through electronic communication media. Though the latter means were employed later, yet the school has been engaged in training since 1928.

Not less important is to highlight the fact that records created today are more prone to decay than old records created 100 years ago or so. The life expectancy of records received today in Archives is less than what was expected in case of old records. It is all happening when Indian conservation is capable of handling the problems independently and for which we are proud of. The question is why it is happening and who is responsible; People or Archives or Government or lack of modern facilities or something else? If thought deeply, it seems that, more or less every one is responsible for it, as the conservation inspite of travelling a long journey was never wide spread and took shape of a movement. It could not happen as the spiritual zeal, never surfaced to attract people. But who had to take initiative to make conservation a movement—whether people or institution or Government? It was people, it is people and it will be people, not individually, but as an organization. The fact is that even after 50 years of independence, India has no any national policy on conservation. There is no legislation regarding writing materials (paper, ink, carbon paper, typewriter ribbon ink, etc.) storage and their safety and security while in U.S.A., the suitable legislation process in connection with use of paper and ink of quality, started to take a stem in 1639[1] and grew up 1895[2].

No doubt, in the last decade of the 20th century (1997), the archival act came into force but that deals with policy of transfer of records to archives after a stipulated period of 25 years. The use of quality materials for creation of records, their scientific preservation in departmental record rooms as well as in repositories did not get representation in the act. Had these facts been made mandatory, the various activities of archieves right from the creation of records, storage, transfer and preservation would have been streamlined and the scenario of conservation being observed today would have been much more better. But ignorance as such in the act, is really disgraceful and apathetic.

The Action Plan for Future

The demand of the time is a National Conservation Policy and the legislation. What should be the principal element defining a policy and legislation is the major question on which conservationists, managers and administrators may differ in their opinion. But what I feel is that whatever policy is chalked out that should ensure conservation of documentary heritage and made it mandatory. It should also touch the economic criterion. For promoting preservation, provision like identification of custodial agencies, their holdings inviting focus on the condition of records in general and endangered records in particular with the promises for immediate aid to save endangered records together with provision for arranging facilities on the minimum requirement basis may figure in the policy. An apex body committee and zonal committees from the experts of National Archives of India, and other organisations like State Archives may be formed so that they may look into various aspects including the production of quality materials conforming I.S.I. standards and production of acid free alkaline buffer materials (docket cover paper, board, cloth, etc.) which are basic requirement of long term conservation. Besides, provision for sharing information among various countries and organizations on the mateials and technique for conservation, research on conservation, co-operation with various organizations including universities and sister organizations, should be made part of National conservation policy, in order to make it more affective.

The conservation as a whole is not a static programme. It requires changes from time to time with the changes of conditions and materials. To match this flexibility, the National policy may have provision to revise the programme suiting the requirement in a specific period of time so that the new problems may be discussed and suitable way out may be found out. The time has come to think seriously over some below mentioned burning problems, which need immediate attention and may be part of policy and legislation.

(1) Location of archival repository whether in capital cities of the country or in places safe from ravages of modern war, needs consideration. The need to have a National Archives for preserving duplicate copies of records is equally important and be mooted thoroughly. It has become inevitable, in view of a danger of atomic war in the sub-continent. National Archives of India is developing microfilm repository at Regional Centre Bhopal, M.P., but the ideal place for a microfilm repository may be Pachmarhi, region, as climate and the environment of the place is conducive for preservation and is centrally located.

(2) The issue of newspaper conservation, pending from more than three decades now need to be urgently sorted out, because they are known for their importance as secondary source of information. The possible steps to be taken in this connection have already been covered in Chapter 4. However, once again it is emphasized that the establishment of a Newspaper Archives in the country is the right step to conserve this wealth tangibly.

(3) Digitization and computerization etc. of information is gaining momentum as an aid of conservation. So far as dissemination of information in distant places, the system is really appreciable but the system from the conservation point of view has been questioned in advance countries. In a state of developing economy, it is the duty of every citizen to spend every paisa very carefully in order to avoid wastage. We must remember that originals even after their duplication, and computerization are must to be preserved as the life expectancy of non-book records excluding microforms is not dependable. As such, digitisation, before it is adopted in large scale as conservation policy, it should be debated for its *pross* and *cons*.

(4) The new records material such as e-mail is becoming official business communication today. So is the case with records produced by fax machines. In country like Australia, courts have accepted e-mail as legitimate source of evidence and it is, therefore, a subject of legal importance. Their permanence and durability and the process of retaining them permanently is yet to be studied in India.

(5) Non-book materials such as microfilms, audio and video cassettes, tapes and gramophone discs, magnetic media, optical discs and photographs are rich sources of information. The restorative programme and techniques to be employed for correcting the damaged wealth of this category has gone a long way in European countries but we are lacking even preventive conservation techniques in respect of these materials. New parameter of temperature (minus 6°C) has recently been set in Germany to preserve colour films. Here we are to conduct researches

on these materials to test their suitability and longevity in Indian environmental conditions so that the right material and technique be adopted for creation, storage, preservation and restoration of this wealth.

(6) Most of the systems adopted initially for conserving documentary wealth have become outdated now. One such case is related to fire detection and suppression. Even carbon-dioxide, which was once most wanted material, is seen unsuitable for suppression of fire today. The new fire detectors such as view system monitored by lasers and system working on the basis of air change have replaced smoke and heat detectors. Dry water rider sprinkler system is seen today in many countries the most effective and suitable fire suppression system for Archives and Libraries. Air washing, change of air in stack areas, smoke expulsion from there, concept of natural air conditioning of Archival buildings, are the most desired activities to be tested and adopted if suits to our requirement for the preservation of records economically.

(7) Computers are working wonders today. They may be suitable tool to monitor environment in repositories. Needs a debate.

The concept of prolonging life expectancy of records for permanent preservation can no way be sidetracked. To achieve it various measures including the activities involved for first phase conservation must be executed without delay. It is only possible when these are made mandatory for which a legislation is a must. It would definitely provide more teeth to the jaws of agencies like National Archives of India in executing the activities and reaping a rich reward.

Obviously to keep pace with the time, a national conservation policy and legislation, is need of the day. This can definitely change the conservation scenario in the Indian sub-continent and in developing and under developed countries.

REFERENCES

+1. "Harold S. Burt, in his article Inks and papers for use in Govt. offices in U.S.A.", published in the *Indian Archives*, Vol. No. 1-2, Jan-Dec. 1956. p. 13, has stated that in 1639, the General Court of

Massachusetts Bay Colony passed an act reading as, "It is ordered and declared that hence forward, every judgement, with all its evidence be recorded in a book to be kept to posterity." Later, in 1816, the necessity for a very definite type of paper was recognized and the following law passed by the legislature of that same state. "All matters which are to be entered of record in any office of public record shall be so entered or recorded on paper made wholly of linen, of firms texture, well glayed & well finished. Nearly one hundred years later, Connecticut placed on its law books the following;

"No person having custody of any book of record or registry in any department or office of the state or any country, city, town, borough or probate district shall use of permit to be used for recording purpose any book which shall not be composed wholly of a standard mill brand paper with dated water mark approved by the examiner of public records." This law has remained substantially the same expect for the addition of a penalty of one hundred dollars fine for failure to obey. This penalty also may be imposed for the failure to use an approved ink.

+2. "Specifying inks for Government offices had its beginning in the United States in 1891 when the Legislature of Massachusetts authorized a standard ink and provided a penalty for failure to use it.

The standard ink of Massachusetts was used as standard by Connecticut in 1895, when the Legislature passed the following Law: "No person having the care or custody of any book of records or registry in any of the departments or offices of the state or of any country, city, town, borough or probate district therein shall use or allow to be used such books any ink excepting such as is furnished by the Secretary of State. The inks so furnished before acceptance by the Secretary shall be examined by a chemist to be designated by the Secretary etc. Soon after this, the federal Government in its treasury Department began ink standardization for use on official records."

APPENDICES

APPENDIX-I

LIST OF PAPER MILLS AS ON 1.1.1987

Category I Mills Having Capacity above 20,000 Tonnes Per Annum		
Sl. No.	*Name of the Mill and Location*	*Capacity (Tonnes)*
1.	Orient Paper Mills (Brijrajnagar Unit), Brijrajnagar, Orissa	76,000
2.	India Paper Pulp Co. Ltd., Hazinagar, 24 Parganas, West Bengal	23,000
3.	Bengal Paper Mills Co. Ltd., Ranigunj, West Bengal	50,000
4.	Star Paper Mills Ltd., Saharanpur, Uttar Pradesh	46,000
5.	Rohtas Industries Ltd., Dalmianagar, Bihar	60,000
6.	Titaghur Paper Mills Co. Ltd., (i) Titaghur; (ii) Kankinara, West Bengal	75,000
7.	Straw Products Ltd., Rayagada, Orissa	50,500
8.	Ballarpur Industries Ltd., (Shree Gopal Division), Yamunanagar, Haryana	58,000
9.	Straw Product Ltd., Bhopal, Madhya Pradesh	25,700
10.	Orient Paper Mills Ltd., Amlai, Madhya Pradesh	85,000
11.	Ballarpur Industries Ltd. (Ballarpur Division), Ballarpur, Maharashtra	80,500

[contd...]

Appendix-I *[contd...]*

Sl. No.	*Name of the Mill and Location*	*Capacity (Tonnes)*
12.	Laxmi Board and Paper Mills (P) Ltd., Kalyan, Maharashtra	22,500
13.	Sirpur Paper Mills Ltd., Sirpur Kaghaznagar, Andhra Pradesh	66,100
14.	Andhra Pradesh Pulp and Paper Mills Ltd., Rajamundry, Andhra Pradesh	75,000
15.	Hindustan Paper Corporation Ltd., Jagi Road, Nowgong, Assam	1,00,000
16.	Mysore Paper Mills Ltd., Bhadravati, Karnataka	37,000
17.	West Coast Paper Mills Ltd., Dandeli, Karnataka	60,000
18.	Punalur Paper Mills Ltd., Punalur, Kerala	33,000
19.	Seshasayee Paper and Boards Ltd., Erode, Tamil Nadu	55,000
20.	Ashok Paper Mills Ltd., Jogighopa, Assam	27,000
21.	Sree Rayalaseema Paper Mills Ltd., Kurnool, Andhra Pradesh	12,000
22.	Bhadrachalam Paperboard Ltd., Bhadrachalam, Andhra Pradesh	50,000
23.	Rohit Pulp and Paper Mills Ltd., Khadki, Udvada, Gujarat	24,600
24.	Nagaland Pulp and Paper Co. Ltd., Tuli, Nagaland	33,000
25.	Titaghur Paper Mills, Chowdwar, Orissa	22,000
26.	Sewa Paper Ltd., Korapur, Orissa	25,000
27.	Tamil Nadu Newsprint and Papers Ltd., Pugalur, Tamil Nadu	40,000
28.	Central Pulp Mills Ltd., Fort Songad, Gujarat	25,500
		13,67,400

Category II
Mills Having Capacity above 10000-20000 Tonnes Per Annum

Sl. No.	*Name of the Mill and Location*	*Capacity (Tonnes)*
1.	Trireni Tissue Ltd., Chandrahati, Hooghly, West Bengal	13,500
2.	Paper and Pulp Conversions Ltd., Khopoli, Maharashtra	20,000
3.	Mandya National Paper Mills Ltd., Belagula, Karnataka	16,500
4.	Sun Paper Mills Co. Ltd., Cheran Mahadevi, Tamil Nadu	15,000
5.	Sehgal Paper Mills Ltd., Daruhera, Haryana	15,000
6.	Ellora Paper Mills Ltd., Tumsar, Maharashtra	13,200
7.	Pudumjee Pulp and Paper Ltd., Thergaon, Chinchwad, Maharashtra	11,000
8.	Wood Paper (Formerly M/s Arvind Board and Paper Mills), Bilimora, Gujarat	14,000
9.	M/s United Pulp and Paper Mills Ltd., Hoshiarpur, Punjab	20,000
10.	Zenith Paper Ltd., Hoshiarpur, Punjab	19,800
11.	Century Pulp and Paper, Lalkua, Nainital, Uttaranchal	20,000
12.	Sri Laxmi Saraswathi Paper Ltd., Nizamabad, Andhra Pradesh	15,500
13.	Neptune Papar Mills Ltd., Midnapur, West Bengal	16,500
14.	Brooke Bond India Ltd., Bilaspur, Madhya Pradesh	15,500
15.	F. Pudumjee & Co. Ltd., Chinchwad, Maharashtra	15,000
16.	Balkrishna Paper Mills Ltd., Ambivli, Maharashtra	14,400
17.	T.T. Cardboard and Paper Mills Ltd., Bhavani Nagar, Tamil Nadu	10,000
18.	Ashok Paper Mills Ltd., Rameshwar Nagar, Bihar	13,500
		2,78,600

Category III Mills Having Capacity up to 10,000 Tonne Per Annum		
Sl. No.	*Name of the Mill and Location*	*Capacity (Tonnes)*
1.	Western India Match Co. Ltd., Alambazar, Calcutta, West Bengal	7,500
2.	East End Paper Industries Ltd., Hooghly, West Bengal	6,000
3.	Universal Paper Mills Ltd., Thargram, West Bengal	10,000
4.	Basant Paper Mills Ltd., Patwa, Uttar Pradesh	5,275
5.	U.P. Straw and Agro Products Ltd., Aghwanpur, Uttar Pradesh	10,000
6.	Gujarat Paper Mills Ltd., Barejadi, Gujarat	6,000
7.	Jayant Paper Mills Ltd., Utran, Gujarat	9,000
8.	Associated Pulp and Paper Mills Ltd., Ahmedabad, Gujarat	9,900
9.	Ajanta Paper and Gen. Products (P) Ltd., Kalyan, Maharashtra	6,600
10.	Pravara Pulp and Paper Mills, Pravaranagar, Maharashtra	7,800
11.	Amravati Sri Venkatesa Paper Mills Ltd., Madathukulam, Tamil Nadu	8,000
12.	Delhi Pulp Industries, Faridabad, Haryana	9,000
13.	Perfect Pack Ltd., Faridabad, Haryana	9,000
14.	Papyrus Papers Ltd., Kalyani, West Bengal	6,000
15.	Pondicherry Papers Ltd., Pondicherry	9,000
16.	Delta Paper Mills Ltd., Bhimavaram, Andhra Pradesh	9,000
17.	Sugam Paper and Board Mills (P) Ltd., Surat, Gujarat	6,000
18.	Nath Pulp and Paper Mills Ltd., Aurangabad, Maharashtra	9,000
19.	Badheka Paper Mfg. Co. (P) Ltd., Nanjangud, Karnataka	9,000
20.	Mukerian Papers Ltd., Mukerian, Punjab	9,000
21.	Sivalik Cellulose Ltd., Gajraula, Uttar Pradesh	9,000
22.	South India Paper Mills (P) Ltd., Nanjangud, Karnataka	6,000
23.	Guardian Paper Ltd., Bamuluru, Andhra Pradesh	10,000
24.	Orient Steel and Wire Ind. (P) Ltd., (Haryana Paper Mills), Faridabad, Haryana	7,500
25.	Patel Paper Mills (P) Ltd., Surat, Gujarat	6,000
26.	Sarvodaya Paper Mills Ltd., Sikandarabad, U.P.	9,000
27.	ABC Paper Mills Ltd., Sailakhurd, Punjab	10,000
28.	Vinod Paper Mills Ltd., Sangrur, Punjab	10,000

[contd...]

Appendix-I Category-III [contd...]

Sl. No.	*Name of the Mill and Location*	*Capacity (Tonnes)*
29.	Saraf Paper Mills, Alwar, Rajasthan	7,500
30.	Sri Venkatesa Paper and Boards (P) Ltd., Madathukulam, Tamil Nadu	6,000
31.	Ansal Paper Mills, Sikandarabad, Uttar Pradesh	10,000
32.	Coastal Paper Ltd., Near Kadiam, Dist. East Godavari, Andhra Pradesh	10,000
33.	M/s Jayakwadi Pulp and Paper Mills Ltd., Ramtek, Maharastra	9,000
34.	North Bihar Sugar Mills Ltd., Champaran, Bihar	7,500
35.	M/s Shiva Paper Mills Ltd., Rampur, Uttar Pradesh	10,000
36.	Shree Padma Paper Mills Ltd., Alwar, Rajasthan	9,000
37.	M/s Kankariya Chemical Industries, Kalol, Gujarat	10,000
38.	M/s Ganapati Pulp and Paper Mills Ltd., Rajeda, Gujarat	10,000
39.	M/s Saiyadri Paper Mills Ltd., Sehore, Madhya Pradesh	6,000
40.	Shreyans Paper Mills Ltd., Malerkotla, Punjab	10,000
41.	Surya Chandra Paper Mills Ltd., East Godavari, Andhra Pradesh	6,000
42.	Circar Paper Mills Ltd., Nellore, Andhra Pradesh	10,000
43.	Nagarjuna Paper Mills Ltd., Pattancheru, Andhra Pradesh	10,000
44.	Wrapaids Ltd., Kollegal, Karnataka	10,000
45.	Shri Vindhya Paper Mills Ltd., Jalgaon, Maharashtra	10,000
46.	Shirke Paper Mills (P) Ltd., Satara, Maharashtra	10,000
47.	Sukhna Paper Mills Ltd., Rajpura, Punjab	10,000
48.	Shree Bhawani Paper Mills Ltd., Rai Bareilly, Uttar Pradesh	10,000
49.	Supreme Paper Mills Ltd., Nadia, West Bengal	10,000
50.	Gangappa Paper Mills Ltd., Panruti, Tamil Nadu	10,000
51.	Pratap Papers Ltd., Gurdaspur, Punjab	10,000
52.	Vamsadhara Paper Mills Ltd., Srikakulam, Andhra Pradesh	7,500
53.	Rollatainers Ltd., Panipat, Haryana	9,000
54.	Pamwi Speciality and Tissue Papers Ltd., Barotiwala, Himachal Pradesh	5,500

[contd...]

Appendix-I Category-III [contd...]

Sl. No.	*Name of the Mill and Location*	*Capacity (Tonnes)*
55.	Porwal Pulp and Paper Mills Ltd., Gadchiroli, Maharashtra	6,000
56.	Shri Datta Shetkari Sahakari Sakhar Karkhana Ltd., Kolhapur, Maharashtra	6,600
57.	Dhanalakshmi Paper Mills (P) Ltd., Madurai Dist., Tamil Nadu	6,000
58.	Sanjay Paper and Chemical Ind. Ltd., Basti, Uttar Pradesh	6,600
59.	Cauvery Paper Ltd., Kollegal, Karnataka	10,000
60.	Rama Pulp and Paper Ltd., Vapi, Gujarat	10,000
61.	Indian Dairy Corporation, Valsala, Gujarat	7,650
62.	Simplex Mills Co. Ltd., Gondia, Maharashtra	9,000
63.	Sangal Paper Ltd., Mawana, Uttar Pradesh	6,000
64.	Emami Paper Mills Ltd., Balasore, Orissa	6,500
65.	Paper Complex, Shree Waran Sahakari Sakhar Karkhana Ltd., Warananagar, Kolhapur, Maharashtra	6,600
66.	Madhya Bharat Papers Ltd., Bilaspur, Madhya Pradesh	10,000
67.	Gaurav Paper Mills, Nagpur, Maharashtra	10,000
68.	P.V.K. Paper Ltd., Basti, Uttar Pradesh	8,250
69.	Vapi Paper Mills Ltd., Vapi, Gujarat	6,000
70.	Aurangabad Paper Mills, Aurangabad, Maharashtra	9,000
71.	Raigad Paper Ltd., Roha, Maharashtra	8,000
72.	Sardar Sahakari Paper Mills Ltd., Gangadhra, Gujarat	5,940
73.	Rajalakshmi Paper Mills Ltd., Maduthukulam, Tamil Nadu	6,000
74.	IPCO Paper Mills Ltd., Thana, Maharashtra	10,000
75.	Aggarwal Paper Mills (P) Ltd., Vapi, Gujarat	7,000
76.	Shri Krishna Paper and Ind. Pvt. Ltd., Bahadurgarh, Haryana	7,000
77.	Indo Afrique Paper Mills Pvt. Ltd., Satara Road, Maharashtra	10,000
78.	Shree Acids and Chemicals Ltd., Moradabad, Uttar Pradesh	10,000
79.	Manaylux Paper and Board Ltd., Basavanguda, Karnataka	10,000
80.	GTC Industries Ltd., Raigad, Maharashtra	7,000
81.	Servall Engineers Ind. Pvt. Ltd., Nailakottai, Tamil Nadu	7,250
82.	The Kopargaon Sahakari Sakhar Karkhana Ltd., Gautam Nagar, Maharashtra	8,250

[contd...]

Appendix-I Category-III [contd...]

Sl. No.	*Name of the Mill and Location*	*Capacity (Tonnes)*
83.	Riverside Paper Mills Pvt. Ltd., Alakpur, Gujarat	10,000
84.	Rahuri Sahakari Sakhar Karkhana Ltd., Rahuri, Maharashtra	8,250
85.	Everest Paper Mills (P) Ltd., Ganganagar, 24 Parganas, West Bengal	3,000
86.	G.T. Paper Mills Ltd., Ujhani, Uttar Pradesh	3,000
87.	Thakur Paper Mills Ltd., Samastipur, Bihar	3,000
88.	Indian Card Board Industries, Howrah, West Bengal	5,000
89.	Rajendra Paper Mills, Faridabad, Haryana	4,000
90.	ACME Paper Mills (Formerly Mandidip Paper Mills), Sehore, Madhya Pradesh	3,000
91.	Sabarmati Paper Udyog (P) Ltd., Vijapur, Gujarat	4,500
92.	Parasuram Paper Mills, Kherdi, Maharashtra	2,100
93.	Deccan Paper Mills Co. Ltd., Mundhwa, Maharashtra	3,600
94.	Premier Paper Mills Ltd., Roha, Maharashtra	3,600
95.	Vidarbha Paper Mills Ltd., Sihora, Maharashtra	2,200
96.	Bombay Pulp and Paper Mfg. Co. Ltd., Dambivali, Maharashtra	4,000
97.	Afsons Industrial Corpn. Ltd., Kalyan Maharashtra	2,400
98.	Senapathy Whiteley (P) Ltd., Ramanagaram, Karnataka	4,920
99.	Grasim Industries Ltd., Birlakootam, Khozhikode, Kerala	2,700
100.	Packart (Pulp and Paper) Ankleshwar, Gujarat	3,500
101.	Kabini Papers Ltd., Nanjangud, Karnataka	3,000
102.	Caprihans India Ltd., Roha, Maharashtra	5,000
103.	Ankleshwar Paper and Board Mills Ankleshwar, Gujarat	3,000
104.	PSD Paper Mills (P) Ltd., Faridabad, Haryana	3,000
105.	Shanti Paper Mills, Sonepat, Haryana	4,500
106.	Valod Antyoday Shakari Paper Mills (P) Ltd., Valod, Gujarat	3,000
107.	Premier Paper and Board Mills (P) Ltd., Kota, Rajasthan	3,000
108.	Murari Paper Mills, Faridabad, Haryana	3,000

[contd...]

Appendix-I Category-III [contd...]

Sl. No.	*Name of the Mill and Location*	*Capacity (Tonnes)*
109.	Dhavalgir Paper Mills (P) Ltd., Gangadhara, Gujarat	4,500
110.	Hari Om Paper Mills (P) Ltd. Gangadhara, Gujarat	4,000
111.	Bagla Paper Mills, Sahibabad, Uttar Pradesh	2,400
112.	Atlas Paper Mills (P) Ltd., Vapi, Gujarat	4,000
113.	Modinagar Paper Mills Ltd., Modinagar, Uttar Pradesh	2,800
114.	Balaji Paper Mills (P) Ltd. Mayuram, Tamil Nadu	3,000
115.	Advance Paper Mills Ltd., Gunjol, Rajasthan	4,500
116.	M/s Raman Boards Ltd., Nanjangud, Karnataka	3,000
117.	Central Indian Spinning, Wvg. Mfg. Co. Ltd., Nagpur, Maharashtra	2,500
118.	Charminar Paper Ltd., Medak, Andhra Pradesh	3,000
119.	Nizam Paper and Board Mills Ltd., Alangudi, Tamil Nadu	5,000
120.	Shree Ambeshwar Paper Mills (P) Ltd., Ankleshwar, Gujarat	3,000
121.	Babri Paper Mills Ltd., Muzaffarnagar, Uttar Pradeh	2,800
122.	Shamli Paper Mills Ltd., Shamli, Uttar Pradesh	2,800
123.	Raval Papers Ltd., Rai Bareilly, Uttar Pradesh	2,640
124.	Surat Board and Paper Mills, Udhna, Gujarat	3,000
125.	Pennar Paper Ltd., Cuddapah, Andhra Pradesh	4,950
126.	Grace Paper Industries (P) Ltd., Bharuch, Gujarat	3,600
127.	Sidharath Papers Ltd., Daruhera, Haryana	2,400
128.	Ruchira Papers Ltd., Kalaamb, Himahal Pardesh	2,310
129.	Ogli Paper (P) Ltd. Kalaamb, Himachal Pradesh	3,000
130.	Barotiwala Paper Mills (P) Ltd. Barotiwala, Himachal Pradesh	3,000
131.	Paper Packaging (P) Ltd. Bangalore, Karnataka	4,950

[contd...]

Appendix-I Category-III [contd...]

Sl. No.	Name of the Mill and Location	Capacity (Tonnes)
132.	Jagotia Paper Mills (P) Ltd. Rupaneli, Rajasthan	3,650
133.	E.C.S. Paper Ltd. Bhiwadi, Rajasthan	3,300
134.	Konark Paper and Industries Ltd., Jharia (Mayurbhanj), Orissa	4,800
135.	Papanasam Paper Mills Ltd., Mela Ambasamudram, Tamil Nadu	2,100
136.	Manjushree Paper Mills (P) Ltd., Saharanpur, Uttar Pradesh	3,000
137.	Nav Bharat Paper Products Modinagar, Uttar Pradesh	3,000
138.	Tissue and Fibres Ltd., Nalagarh, Himachal Pradesh	4,500
139.	Sandeep Paper Mills (P) Ltd., Ghaziabad, Uttar Pradesh	3,000
140.	Visalakshi Mills (P) Ltd., (Paper Division), East Godavari, Andhra Pradesh	4,200
141.	South Delhi Fabricator (I) (P) Ltd., Vidisha, Madhya Pradesh	3,300
142.	Swati Paper Mills (P) Ltd., Kotkapura, Punjab	2,500
143.	Khurana Paper Mills (P) Ltd., Makhiyali, Uttar Pradesh	3,000
144.	M.K. Paper Board Mills (P) Ltd., Ropar Punjab	3,000
145.	Abhyudhya Paper Mills Ltd., Faizabad, Uttar Pradesh	3,300
146.	Mayur Paper Mills Ltd., Chandrapur, Maharashtra	3,000
147.	Aristo Pulp and Paper Ltd., Bundi, Rajasthan	4,500
148.	Satia Paper Mills Ltd. Muktsar, Punjab	4,950
149.	Baruani Paper Industries Ltd., Tilrathi, Bihar	2,250
150.	Lalji Paper Mills (P) Ltd., Barotiwala, Himachal Pradesh	3,300
151.	Kanwar Paper Mills (P) Ltd., Kala Amb, Himachal Pradesh	3,100
152.	Gupta Paper Mills (P) Ltd., Sahibabad, Uttar Pradesh	3,000
153.	Aliyar Paper Mills (P) Ltd., Pollachi, Tamil Nadu	5,000
154.	Globe Paper Mills Ltd., Burdwan, West Bengal	2,240
155.	Panchsheel Paper Mills Ltd., Jalgaon, Maharashtra	3,000
156.	Village Paper (P) Ltd., Una, Himachal Pradesh	3,000
157.	Trilok Paper (P) Ltd., Nandej, Gujarat	3,000
158.	Cheema Paper Mills (P) Ltd., Nainital, Uttar Pradesh	4,950
159.	Kalika Board and Paper Mills (P) Ltd., Hapur, Uttar Pradesh	2,500
160.	Satsuwan Paper Mills Ltd., Alwar, Rajasthan	4,400

[contd...]

Appendix-I Category-III [contd...]

Sl. No.	Name of the Mill and Location	Capacity (Tonnes)
161.	Hapuria Paper Mills (P) Ltd., Sahibabad, Uttar Pradesh	2,400
162.	Adam Paper Products (P) Ltd., Paontasahib, Himachal Pradesh	3,600
163.	Sakambri Paper Mills Badaun, Uttar Pradesh	3,300
164.	Chemo Pulp Tissues Ltd., Ghaziabad, Uttar Pradesh	3,300
165.	B.K. Paper (P) Ltd., Kalyan, Maharashtra	3,500
166.	Agrashakti Paper Mills Ltd., Vapi, Gujarat	2,800
167.	Ruby Paper Mills (P) Ltd., Vapi, Gujarat	2,700
168.	Haria Paper Mills (P) Ltd., Vapi, Gujarat	2,800
169.	Manish Paper and Board Mills Ltd., Sahibabad, Uttar Pradesh	4,500
170.	Shivashakti Paper Mills Ltd., Visisha, Madhya Pradesh	3,000
171.	Shell Paper (P) Ltd., Barotiwala, Himachal Pradesh	3,000
172.	Delux Board and Paper Mills, Vapi, Gujarat	2,880
173.	Yash Paper Ltd., Faizabad, Uttar Pradesh	3,010
174.	Nathani Paper Mills Ltd., Vapi, Gujarat	3,500
175.	Agarwal Pulp and Paper Mills, Muzaffarnagar, Uttar Pradesh	3,000
176.	Bansi Lal Paper Mills (P) Ltd., Bhiwadi, Rajasthan	3,500
177.	R.K. Mills Board (P) Ltd., Muzaffarnagar, Uttar Pradesh	3,300
178.	GVG Paper Mills (P) Ltd., Anna Dt. Tamil Nadu	3,000
179.	Sikka Paper and Sugar Mills Ltd., Muzaffarnagar, Uttar Pradesh	2,640
180.	Satpur Paper and Sugar Mills Ltd., Godarwara, Madhya Pradesh	3,000
181.	Vivardhan Paper Mills Ltd., Madathukulam, Tamil Nadu	5,000
182.	Shree Paper Mills, Samalkot, Andhra Pradesh	4,000
183.	Shri Gopinath Paper Mills Barranagre, West Bengal	1,080
184.	Priti Paper Board Mills (P) Ltd., Hooghly, West Bengal	1,200
185.	P.G. Paper Mills Co. (P) Ltd., Dum Dum, West Bengal	1,500
186.	Brhmani Paper Mills (P) Ltd., Ghaziabad, Uttar Pradesh	1,440
187.	East Coast Paper and Products (P) Ltd., Howarah, West Bengal	1,500
188.	Swatantra Bharat Paper Mills,	

[contd...]

Appendix-I Category-III [contd...]

Sl. No.	*Name of the Mill and Location*	*Capacity (Tonnes)*
	Ghaziabad, Uttar Pradesh	2,000
189.	Gopal Paper and Board Mills, Ghaziabad, Uttar Pradesh	1,500
190.	Aknos Paper Mills (P) Ltd., Ghaziabad, Uttar Pradesh	1,500
191.	Alok Paper Industries, Indore, Madhya Pradesh	1,200
192.	Steadfast Paper Mills Ltd., Gungra, Gujarat	1,050
193.	Parekh Paper Mills, Gondal, Gujarat	1,000
194.	Speciality Paper Mills Ltd., Vapi, Gujarat	1,440
195.	Upper India Couper Paper Mills Ltd., Lucknow, Uttar Pradesh	2,000
196.	Providence Paper Mills Ltd., Santacruz, Bombay	510
197.	Kamrup Paper Mills, Amingaon, Assam	1,000
198.	Venus Paper Mills, Faridabad, Haryana	2,000
199.	Annapurna Paper Mills, Kangeri, Karnataka	700
200.	Agarwal Paper Mills, Muzaffarnagar, Uttar Pradesh	2,000
201.	Saurashtra Paper and Board (P) Ltd., Rajkot, Gujarat	1,875
202.	Shree Lakshmi Narayan Paper, Nanded, Maharashtra	1,650
203.	Suri Paper and Chemicals (P) Ltd., Daruhera, Haryana	1,500
204.	Bhikusa Paper (P) Ltd., Nasik, Maharashtra	1,500
205.	Ranisati Paper Mills (P) Ltd., Ghaziabad, Uttar Pradesh	1,500
206.	Ganesh Paper Mills, Ghaziabad, Uttar Pradesh	1,825
207.	Chandigarh Paper and Board Mills (P) Ltd., 26, Industrial Area, Chandigarh	1,500
208.	Dhampur Sugar Mills Ltd., (Pragati Paper Mills), Shampur, Uttar Pradesh	1,500
209.	Sri Ram Paper Mills (P) Ltd., Laskana, Gujarat	1,200
210.	Super Paper Mills (P) Ltd. Muzaffarnagar, Uttar Pradesh	1,800
211.	Shri Bankey Bihari Board Mills, Ghaziabad, Uttar Pradesh,	900
212.	Rajsheel Paper (P) Ltd., Nanjangud, Karnataka	1,500
213.	Kododra Paper Industries (P) Ltd., Kododra, Gujarat	1,800

[contd...]

Appendix-I Category-III [contd...]

Sl. No.	*Name of the Mill and Location*	*Capacity (Tonnes)*
214.	Airofil Paper Ltd., Banmore, Madhya Pradesh	1,650
215.	Bhargava Paper (P) Ltd., Dadri, Uttar Pradesh	1,650
216.	Mahendra Pulp and Paper Mills (P) Ltd., Vapi, Gujarat	1,500
217.	Satya Paper Mills, Nangli, Punjab	1,800
218.	Arihant Paper (P) Ltd., Sikandarabad, Uttar Pradesh	1,500
219.	Rajesh Paper Mills Ltd., Shikohabad, Uttar Pradesh	2,000
220.	Jagdamba Paper Ind. (P) Ltd., Sirsa, Haryana	1,950
221.	Kulik Paper Ind. (P) Ltd., Raiganj, West Bengal	1,440
222.	Singhal Paper (P) Ltd., Ghaziabad, Uttar Pradesh	1,200
223.	B.K. Paper Mills (P) Ltd., Bhatian, Punjab	1,200
224.	Govardhan Kagaj Udyog, Agra, Uttar Pradesh	1,200
225.	J.K. Paper Mills (P) Ltd., Daruhera, Haryana	1,980
226.	Bishwanath Paper Mills (P) Ltd., Purnea, Bihar	1,350
227.	National Paper Mill, Bulsar, Gujarat	2,000
228.	Sarao Paper and Allied Product (P) Ltd., Barotiwala, Himachal Pradesh	2,000
229.	Roshan Lal Paper Mills (P) Ltd., Bhatinda, Punjab	1,980
230.	Kunniyar Paper and Board Mills Ltd., Thanjavur, Tamil Nadu	1,485
231.	Charan Kamal Cardboard Paper Mills (P) Ltd., Sangramasahib, Punjab	1,500
232.	M.P. Paper & Plastic India (P) Ltd., Vidisha, Madhya Pradesh	1,000
233.	Nagpur Paper Mills (P) Ltd., Manegaon, Maharashtra	1,200
234.	Ameeta Paper (P) Ltd., Raigad, Maharashtra	2,000
235.	Urvashi Pulp and Paper Mills Ltd., Ankleshwar, Gujarat	1,344
236.	Aryabhatt Paper Mills (P) Ltd., Darbhanga, Bihar	1,800
237.	Unitech Paper Mills (P) Ltd., Midnapur, West Bengal	1,650
238.	Three Star Paper Mills (P) Ltd., Dadri, Utar Pradesh	1,650
239.	Gayatri Paper Mills Pvt. Ltd., Raigad, Maharashtra	1,500
	Western India Paper Board Mills (P) Ltd., Vikroli, Maharashtra	1,800
240.	Sangamner Bhag Sahakari Shakar Karkhana, Amrutnagar, Maharashtra	1,800
		11,20,004

APPENDIX-II

RAW MATERIALS FOR PAPER AND BOARD

+2	*Woods/Bamboo*	*Indian Names*	*Botanical Names*
1.	Woods	Silver Fir or rai	Abies Pindrow
2.	–do–	Spruce	Picca Morinda
3.	–do–	Chir	Pinus-Lonifoliasaplings
4.	–do–	Paper Mulberry	Broussonetia Papyrifera
5.	–do–	Pula or Bends	Kydia Calycina
6.	–do–	Gengwa	Excaecaria-agallocha
7.	–do–	Jhingan	Odina wodier
8.	–do–	Kharpat	Garuga Pinnata
9.	–do–	Salai	Boswellia Serrata
10.	–do–	Bluegum	Eucalyptus globurus
11.	–do–	Chir (Mature Wood)	Pinus longifolia
12.	Bamboo	Bamboo	Bambusa nutans
13.	–do–	–do–	Dendro Calamus strictus
14.	–do–	–do–	Bambusa tulda

APPENDIX-III

NOMENCLATURE OF PAPER AND PAPERBOARD LAID DOWN BY THE DIRECTORATE GENERAL OF TECHNICAL DEVELOPMENT

I. Printing Paper:

1. White printing paper
2. Coloured printing paper
3. Imitation art paper
4. Offset printing paper
5. Litho-printing paper (including maplitho)
6. Duplicating and antique paper
7. Unbleached printing paper
8. MG Poster (white/coloured) paper
9. Coated (art and chrome) paper

II. Writing Papers:

1. Account-book paper
2. Azurelaid/ledger paper
3. Typewriting paper
4. Cream laid wove paper
5. Airmail/manifold paper
6. Cheque paper
7. Bond Paper
8. Drawing paper/cartridge paper

III Packing and Wrapping Papers:

1. Sack kraft paper
2. Kraft paper

3. Manilla/cover paper
4. Match paper
5. Common wrapping paper
6. Tissue paper
7. Grease-proof/glassine/parchment paper

IV. Other varieties:

1. Cigarette paper
2. Blotting paper
3. Filter paper
4. Base paper (for various purposes)
5. Crepe paper
6. Electrical grade paper
7. Permanent record paper

V. Paperboards:

1. Coated (art and chrome) board
2. Black centered board
3. Duplex/triplex board
4. Ticket board
5. Grey board
6. Mill board
7. Pulpboard/boboard
8. Postcard board
9. Presspahn (electrical and industrial grades)
10. Vulcanised fibre sheet
11. Tag board

APPENDIX-IV

DATA RELATED TO EFFECT OF SULPH-ARSENIC OR PAPER

Sample No.	*Particulars of Sample*	*Treatment*	*Tensile Strength in Ib.*	*Average Tensile strength*	*Folding Endur-ance*	*Average folding endur-ance*	*Bursting strength Ib./sq.in.*	*Average bursting strength in Ib./sq.in.*	*Loss of tensile strength %*	*Loss of folding endur-ance %*	*Loss of burst-ing stren-gth%*
M-1	Handmade paper 28" x 20" weight 26.5 Ib. per 500 sheets. machine direction	Untreated	14.00 15.50 14.75	14.75	4935 1993	4919	62 64 64 57 42 63 50 60	57.8	–	–	–
M-2	Ditto	Accelerated ageing	13.50 12.75 12.75	13.08	3839 3195	3817	–	–	11.32	22.41 (35.72*)	–
M-3	Ditto	Treated with sulph-arsenic solution	13.00 12.75 12.75	12.66	3184 3176	3180	40 50 45 45 44 55 49 59	47.3	14.17	35.35	18.17

[contd.]

Appendix-IV *[contd...]*

Sample No.	*Particulars of Sample*	*Treatment*	*Tensile Strength in lb.*	*Average Tensile strength*	*Folding Endurance*	*Average folding endurance*	*Bursting strength lb./sq.in.*	*Average bursting strength in lb./sq.in.*	*Loss of tensile strength %*	*Loss of folding endurance %*	*Loss of bursting strength%*
M-4	Ditto	Treated with Sulpharsenic and then subjected to accelerated ageing	12.25 13.25 12.00	12.50	3140 2866 2866	3003	–	–	15.25	38.94 (40.54)*	–
C-1	Handmade 28" x 20" weight 26.5 lb. per 500 sheets Cross direction	untreated	–	–	3429 3390	3410	–	–	–	–	–
C-2	Ditto	Accelerated ageing	–	–	2597	2513	–	–	–	26.16	–
C-3	Ditto	Treated with Sulpharsenic solution	–	–	2041 2316	2179	–	–	–	36.08	–
C-4	Ditto	Treated with Sulpharsenic and then subjected to accelerated ageing.	–	–	1993 1954	1974	–	–	–	42.11	–

* Average of machine and cross direction.

APPENDIX-V

FUMIGATION OF BOOK-INSECTS WITH NAPHTHALENE

Species	*under test*	*Number in test vessel*	*Concentation of fumigant in test vessel*	*Time for inactivity*	*Time for paralysis*	*Remarks*
	Test	Six	0.695 gm./litre	Four cockroaches were found to be be inactive after	90-100 minutes	As soon as the fumigant was introduced in the test vessel the insects showed increased activity. Paralysed insects were kept exposed to fresh air for 3 to 4 hours. No sings of activity were shown.
Cock-roach	Control	Six	nil	–	–	Insects in the control vessel were examined next day (20 hours) and all were found active.
	Test	Four	0.695 gm./litre	All the insects showed inactivity after 40-45 minutes	90-100 minutes	Paralysed insects were kept exposed of fresh air. No signs of activity were observed even after 3-4 hours.
Silverfish	Control	Four	Nil	–	–	Insects in the control vessel were examined next day (20 hours) and all were found active.
	Test	Eight	0.695 gm./liters	No observation could be made	90-100 minutes (only two larvae showed signs of inactivity on dis-turbance)	None of the eight larvae showed any signs of activity when left exposed to fresh air for 3-4 hours.
Gastrallus Indian Larvae	Control	Eight	Nil	–	–	Six larvae were found active when examined after 20 hours.

APPENDIX-VI

LAYOUT PLAN FOR MODEL RECORD ROOM

Guidelines on the layout Plan for Model Record Room are as under:

(1) Composition

A record room should be composed of three units i.e., stack area, room for staff and area for attending minor repair and work relating to sterilisation of records.

(2) Location

(i) It should be located preferably on the ground floor, far from any dranage and electric power plant.

(ii) Plinth level of the Record Room should be higher than the surrounding area.

(iii) There should be no active water tap in the Record Room and near the close vicinity of it.

(3) Size of Stack Area & other Units

Size of the stack area should be selected according to the bulk of records as under:

(i) For housing 2,25000 files, a stack having area of 200 sq.m. would be suffice. Suitable dimension may be 20 x 10 m.

(ii) For accommodating files upto 1,50,000, a room having area of 140 sq. m. would be enough. A suitable dimension may be 14 x 10 m.

(iii) For accommodating files up to 50,000, area of stack room should be 72 sq. m. A suitable dimension may be 9 x 8 m.

(iv) The size suggested above, are, suitable to accommodate the records employing double entry shelving system, having provision of gang ways (.70m), Main Gangway (1.40m) and .30 cm. gap between rack and all along the side walls. However, less space would be required if compactus shelving is used.

(v) Room for staff should be selected according to the strength of staff, at the rate of 5 sq.m. per official. However, the size of room to carry out minor repair work and sterilization of records should be approximately 25 sq.m.

Requirement of a Stack Room

General

(i) It should have adequate number of windows promoting free air circulation in the room through cross ventilation.

(ii) Every window should be fitted with metal grills, wire mesh, tinted glasses and water shades.

(iii) Windows should be draped with thick cottom cloth curtain of either colours, red, green, blue, yellow.

(iv) Windows in large number preferably should not face east and west.

(v) Doors should be fitted with automatic push-pull system.

(vi) Outlets of the room should be fitted with wire mesh.

(vii) Room should be damp proof as well as termite proof and provisioned with exit door (s).

(viii) The room should be Fire Proof which can, to some extent, be achieved by adopting

  (a) Copper wiring through conduets.

  (b) Installing main control switches of the area located outside the room.

  (c) Employing additional automatic dampers in centrally air conditioned rooms.

  (d) Sectional wiring system in the rooms having area more than 150 sq. m.

(ix) Room should be fitted with automatic fire alarm detection system (preferably smoke detectors).

(x) Carbon dioxide type fire extenguishes should be installed at convenient points of the room in order to combat the fire if it outbreaks.

(xi) Use of loose wires and additional electrical appliances in the circuit should be avoided.

(xii) Smoking in the area should be prohibited.

Lighting

From the view point of preservation of records, the proper illumination of stack area is essential. For the purpose, concealed diffused light system is useful. Other systems such as smoothering of walls and use of titanium white are effective to reach proper illumination. The light levels in the area should not go beyond 75 lux.

Shelving

(i) The racks preferably be made of rust free iron should be used. The use of wooden racks in general be avoided.

(ii) The shelving system (double entry) should have the following provision.

(a) Gang ways between two successive rows (0.70m)

(b) Main gangway in stack room larger than 100 sq. m. should have a width of 1.40 m.

(c) A gap of 30cm between the frame top of rack and ceiling should be maintained for free air circulation.

(d) A gap of 30cm between rack and wall should be maintained all along the side walls.

(e) For higher density storage and also to protect the records from dust, dirt and light compactus shelving system is the demand of time.

Storage Environment

(i) Stack area should be dust free. For removing dust use of vacuum cleaners is more suitable.

(ii) Temperature range should be 20-25°C.

(iii) Humidity range between 50-60%.

(iv) To control temperature and humidity at optimum range, air conditioning is the right answer.

(v) In non-airconditioned areas however, no stagnent pocket of air should be allowed to form in the stack area. It may be avoided by circulating air using air circulators and maintaining cross ventilation.

(vi) Any cracks and holes on the floor and walls should immediately be repaired and filled with concrete.

(vii) No edibles should be allowed in the stack areas.

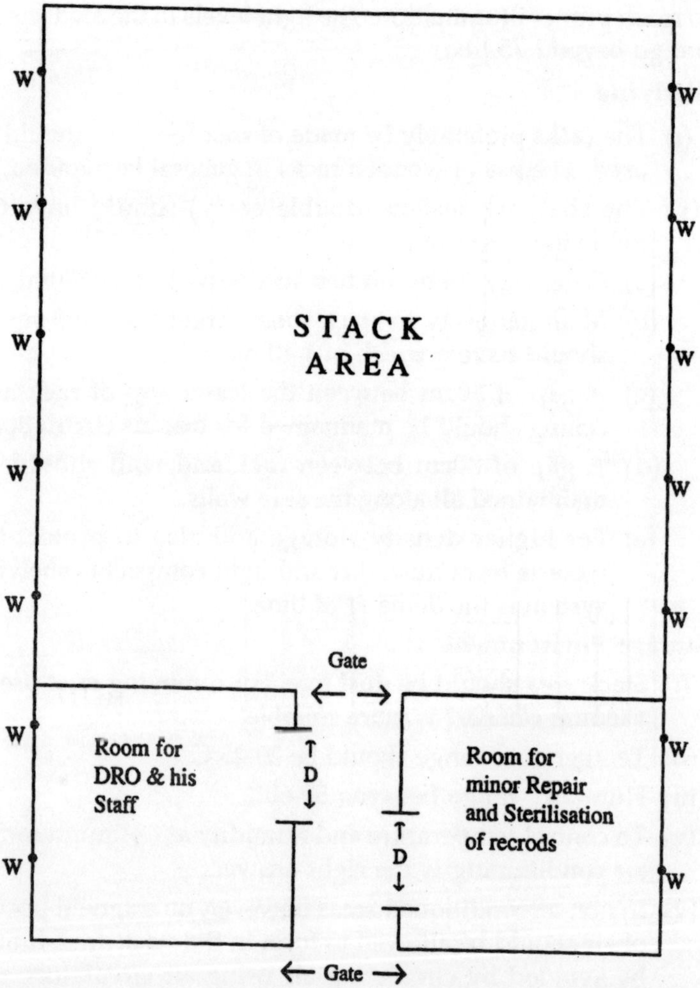

Fig. App. iv.1 : Showing One of the Ways of Composition of Record Room.

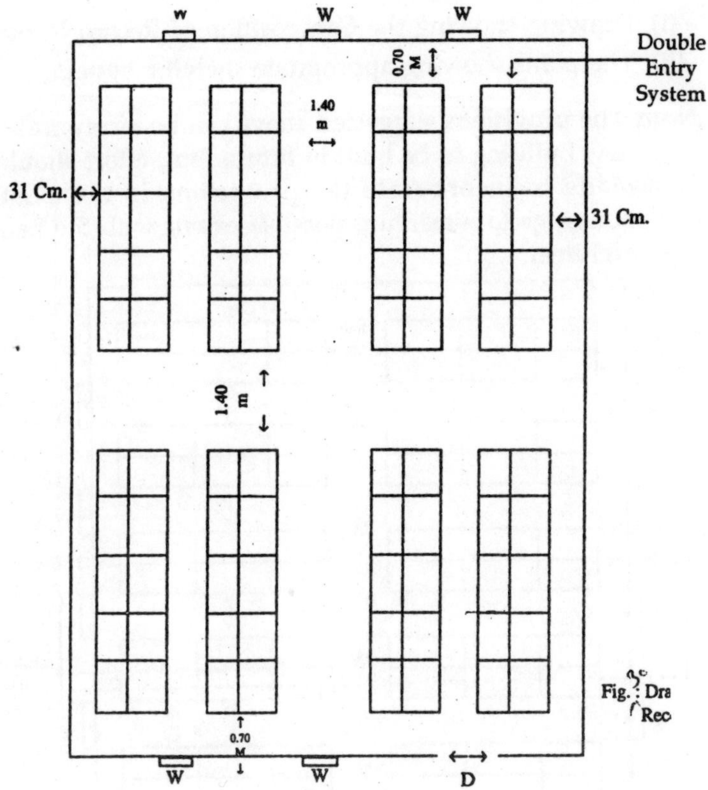

W= Window fitted with metal grills and wire mesh

D= Door Automatic

Fig. App. iv.2 : Drawing showing shelving plan for Departmental Record Room.

(viii) The stack area should be disinfected from time to time specially in the transition periods using pip insecticides and 10% solution of tymol in methanol, alternatively.

(ix) Naphthalene bricks may be used as insect repellents in the areas. (non-conditioned)

Drawings

(i) Drawing showing the Composition of Record Room.

(ii) The plans showing appropriate shelving system.

Note: The provisions suggested above can be incorporated in new building to be built in future. But, effort should be made to incorporate the provisions in the existing buildings to maximum possible extent with the help of architect.

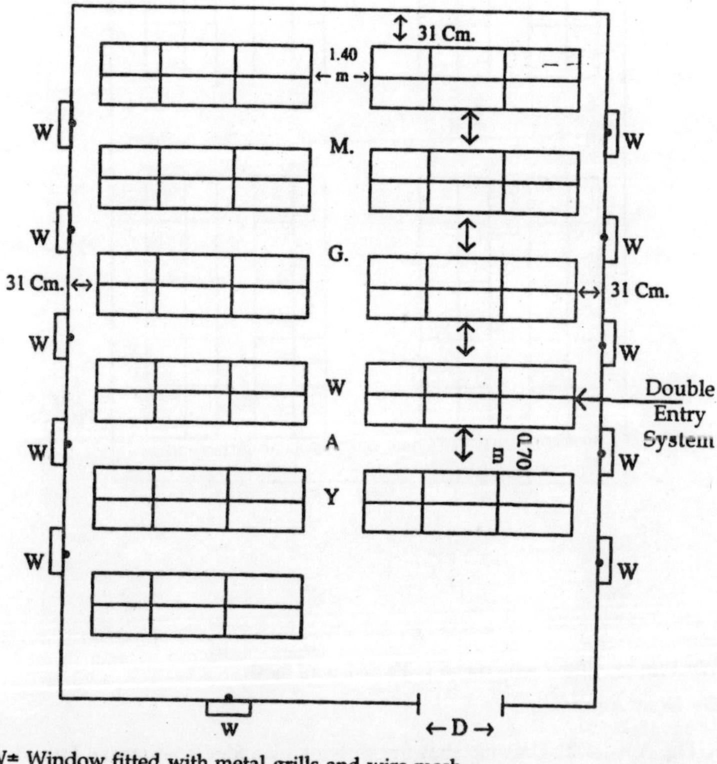

W= Window fitted with metal grills and wire mesh
D= Door Automatic

Fig. App. iv.3 : Showing Shelving plan for Department Record Room.

APPENDIX-VII

LIST OF SOME FIBRES IN PAPER MAKING

Local (Indian) Name	*Parts Used*	*Botanical Name*
Ananas	Leaf fibre	Ananas Sativa
Bamboo, Bans	Tissues, baskets and mats	Bamboo
Bhendi	Stem	Abelmoscus esculentus
Dhak, Palas	Inner bark	Butia frondoso
Madar	Bast fibre	Calotropis Procera
Jute hemp	Bast fibre, gunny bags and waste	Corchorus capsularis
Sun (local Hindi name in Bundelkhand)	Bast fibres, ropes bags etc.	Crotalaria junced
Adarak (ginger)	Whole plant	curcuma varities
Shahtoot white (White mulberry)	Inner barle of young plams	Broussonetia papyrifera
Motha, Mutha	Whole grass	cyperus rotundus,
Ketki, ghaipat	fibre of leaves-waste	Agave ameriana
Hemp	Bast fibre	Cannobus sativa,
Moonj or Sara grass	Leaves, ropes	Saccharum munja
Ketki Kevada	Leaf fibre	Pandanus odoratissimus
Kapas (Cotton)	rags-textiles	Gossypium
Alasi-Javas	Bast fibre	Limum usitatissimum
Kela (Banana)	Stem fibre	Musa vaneties
Bhoosa (straw) wheat	wholestraw	Triticum sativa,
Bhoosa (straw) dhann (Rice)	Whole straw	aryza sativa
Mothi	Bast fibre	Grarddinia heteraphyll
Tal, Palms	Stalks, Leaves	Palae Vanities
Ganna (Sugarcane)	Bagasse	Sacharum officimum

ANNEXURES

ANNEXURE-I

CONSERVATION OF ILLUSTRATED MANUSCRIPTS ON PAPER SUPPORT

The conservation, specially restorative part of illustrated manuscripts poses serious problem due to its mixed structure and involving use of paints. What is additional attached to an illustrated manuscript than an ordinary manuscript is the art work making use of various colours which abit differ from inks. In simple terms, an illustrated manuscript, involves the value of a manuscript and a painting as well. It is not merely to be read but to be viewed also in the way artist intended to be viewed.

A document or manuscript bears only layer of ink employed for depiction which is partially absorbed by the paper carrier. But in case of painting which is technically a stratified structure of one layer over the other and the paint layer, even after absorption by the carrier, the thickness of the portion is perceptible comparative to the manuscript. This is the basic physical difference between the written and painted portion of an illustrated Manuscript.

Two type of illustrated Manuscripts are very common getting places in Museums and Archives in India. Execution of thematic painted work and written contents either on obverse and reverse or on one side only makes one form while another form have art work all around the written contents see plate I. Illustrated kalpasutra, Bhagwatgita (1880), transcribed during the time of Maharaja Ranbir Singh of Kashmir, letter of Raja Khalita Singh of Etam of Bencoolin to Lord Hastings (1819) and a Congratulatary letter by Rai Narain Balli sent to the

Nawab Hamid and Alikhan of Rampur on the occassion of his coronation (1844) are few illustrated manuscripts from the collections of National Archives of India. Flowers, and motifs, painted in red, yellow, black, green and golden colours are the work of art to make the letters, attractive and beautiful. Besides, some illustrated manuscripts, are available in previous collections of National Museum. One, such manuscript (ACC No. 51.250) belonging to the early period of 17th century known for its mix technique of Rajput & Mughal style is the known example of the category.

Problems and Soluitons

All the elements discussed in Chapter four are considered responsible for decay, damage and spoilage of this wealth. The most important fact to be pointed out in case of illustrated manuscripts, irrespective of their class to which they belong, is that some colours, employed for art and painting works such as lead oxide, react with hydrogen sulphide present in atmosphere and cause black blemishes. Light of higher intensity not only disintegrates the support but also bring about fading of colours. Humidity whereas affécts the physical strength of paper adversely, brings about a remarkable change in the refractive index of colours. This change in refractive index, affects the view of the art work, Shade of colours employed bythe artist, has a very meaningful role to play on conveying the message to the viewers. The change of shade is governed by the change in refractive index which is affected in excessive of water and solvents absorbed by both, the support and the paints.

Charring of paper, loss of borders, fading of colours, development of black and brown patches on the paintings mostly employing green pigments (Mughal and Rajput Manuscripts), growth of fungus, stitcking of sheets and perforation caused by insects are the common problems affecting illustrated manuscripts. To counteract the agencies responsible for the problems mentioned above, ideal storage conditons, suitable storage system and good houe keeping are dependable and time tested techniques. The decayed and damaged illustrated manuscripts, however, could be restored and imparted physical strength using following methods.

Restoration

Imparting of strength to the decayed manuscripts and correcting of scars formed due to damages, are the main objectives behind restoration of manuscripts. For the purposes: (i) Solvent lamination employing cellulose acetate foil, acetone and tissue paper as well as chiffon and cellulose acetate foil, (ii) Mounting with chiffon using strach paste, and (iii) Lamination with plastic film+ (cellulose tri-acetate) coated with an adhesive can be employed. However, for brilliant results, the work should be performed by skilled restorers.

Before manuscript is subjected for reinforcement that should be cleaned off unwanted foreign material and deacidefied using suitable techniques. In no case, the painted portion of the manuscript should be allowed to be altered during chemical treatment and restoration. Sometimes chemical cleaning of mss is essential before restoration.

Plastic film +1-This type of quality plastic film is supplied by Marane Plastic Co. of U.K.

ANNEXURE-II

COST-COMPARISON OF VARIOUS REINFORCEMENT TECHNIQUES

Conservation of documentary heritage is a costly affair. Archival Institutions penny earn and pound spend. Because of the reason, these organisations, even in developed countries, face hardship of funds to carry out their conservation programmes smoothly. It is the cost factor alone because of which it is advocated that process of reinforcement should be cheap and the best. In other words, the process should be such that which ensures best results at the minimum cost. But the question is how one can compare the cost involved for various reinforcement techniques such as tissue repair, full pasting (Backing), lamination or chiffon repair etc. Rather, it is difficult to work out cost precisely in some cases, but it can be done if efforted. For the purpose, below mentioned formula, can be used.

Evaluation of cost for reinforcing a document = sum of the cost involved in different elements from (I to VII) below.

(i)	Cost of material employed for reinforcement	+(ii)	Wages paid to the craftmen attending reinforcement work.	+(iii)	Fraction of rent paid for accommodation where work is performed.
+(iv)	Fraction of invested money for purchase of equipment.	+(v)	Electricity charges	+(vi)	Other charges if any.
+(vii)	30% extra charges				

How these all the five factor are involved in restoring a document and cost working out procedure could clearly be understood from a typical example illustrated below.

Example: Machine Lamination of an A4 size document.

To work out the total cost involvement, it is essential to find out the share of mount spent for laminating a document by machine.

(1) Cost of Material Employed for Machine Lamination: The materials required for the purpose of laminating a document are acetate foil, Tissue Paper, Cotton Wool, Tafflon cloth and Acetone. For working out the cost collection of following data is necessary.

(i) Cost of tissue paper per ream with size 50x76 cm– Presume Rs. 2500/- ream.

(ii) Cost of Acetate foil per ream with size 51 x 76 cm– Presume Rs. 5000/ream.

(iii) Cost of Acetone per litre-Presume Rs. 100/- lit.

(iv) Cost of cotton wool - Presume Rs. 40/- bundle of 400 gm.

Besides other subsidiary data required for the purpose is, the number of envelopes prepared in one litre of acetone and one bundle of cotton wool. Cost and consumption of Tafflon Cloth per lot for laminating sheets.

One ream is consisted of 500 sheets of paper. Size of paper and foil will help in determining how many sheets could be lamined in one sheet. The size which has been indicated above is adequate to laminate two documents of A4 size conveniently. Obviously ½ sheet of both, tissue paper and acetate foil, would be required for laminating one document and cost of these can be found out as following.

$$\text{Cost of } \frac{1}{2} \text{ sheet of tissue paper} = \frac{\text{Cost of one ream of Paper}}{\text{Number of sheets in one ream}} \times \frac{1}{2}$$

$$= \frac{2500}{500} \times \frac{1}{2} = \text{Rs. } \frac{5}{2} = \text{Rs. } 2.50$$

$$\text{Cost of } \frac{1}{2} \text{ sheet of Acetate foil} = \frac{5000}{500} \times \frac{1}{2} = \text{Rs. } 5.00$$

Before subjecting the documents for lamination in machine, their envelops are prepared and for this purpose acetone and cotton is used. Presuming, one bundle of cotton wool and one liter of acetone is sufficient to prepare 250 envelops each containing 4 documents. The cost of both per document is worked out as under:

$$\text{Cost of Cotton wool per sheet} = \frac{\text{Cost of Cotton per bundle}}{\text{Number of envelops}} \times \frac{1}{4}$$

$$= \frac{40}{250} \times \frac{1}{4} = \text{Rs. } 0.04 \text{ or Paise } 4$$

$$\text{Cost of acetone per sheet} = \frac{\text{Rs. } 100}{\text{Rs. } 250} \times \frac{1}{4} = \text{Rs. } 0.10 \text{ or Ten Paise}$$

Likewise the cost of Taflon cloth per sheet can be found out using formula

$$= \frac{\text{Cost of Two sheets of Taflon cloth of required size}}{\text{Number of envelops laminated before they finish}} \times \frac{1}{4}$$

$$= \frac{800}{10000} \times \frac{1}{4} = .02 = \text{Rs. } 0.02 \text{ Paise } 2.$$

Cost of material employed for lamination = 2.50 + 5.00 + 0.04 + 0.10 + 0.02 = Rs. 7.66

(2) **Labour charges per sheet :** To workout labour charges the following information is essential.

(i) The salary of craftsman attending work - Presume it is Rs. 6000/ month.

(ii) The envelopes prepared per day per man - Presume 40 envelopes.

(iii) Number of sheets laminated per day-Presume 200 sheets or 50 envelops per head.

(iv) Working days in week- Presume 5 days.

Normally, one person prepares 40 envelops in a day and each envelop contain 4 A4 size documents. In one day 200 documents or 50 envelops at least per head are laminated. A craftman attending the work, will prepare 100 envelops in 2½ days while he will laminate them in two days. In a block of 4½ days, 100 envelops are prepared and laminated by a man. In a month, 20-22 working days are available if it is five day week.

The labour charges per document can be found out applying formula;

$$\frac{\text{Salary per month}}{\text{Number of working days}} \times \begin{array}{c}\text{Number of days in which 100}\\ \text{envelops are prepared and laminated}\end{array} \times \frac{1}{400}$$

$$= \frac{6000}{20} \times 4\frac{1}{2} \times \frac{1}{400}$$

$$= \frac{6000}{20} \times \frac{9}{2} \times \frac{1}{400} = \frac{27}{8} = \text{Rs. } 3.38 = 3.40$$

(3) Fraction of Rent Paid per Document: Presume the rent of the workshop where restoration work is carried out is Rs. 5000/- per month. Three persons are working. One is engaged for lamination, and remaining two for tissue repair and full pasting etc. As per norms, approximately 2000 sheets are laminated (in 4½ days - 400 and in five weeks, 400 x 5 = 2000). Similarly 1200 sheets and 2000 sheets are reinforced by fulpasting and tissue repair technique in a month per head, as per norms.

The total number of sheets reinforced in a month are

2000 + 1200 + 2000 = 5200 Sheets.

$$\text{Rental Charges per sheet} = \frac{\text{Ret paid per month}}{\text{Total sheets repaired in a month}} = \frac{5000/-}{5200} = \text{Rs. } 0.95$$

(4) The Amount invested for procurement of Machine also makes a share in lamination of documents. The cost of machine, its life expectancy and interest on the principal invested for purchase of machine have due share in lamination per sheet.

(i) The cost of Laminator is Rs. 10,00000/-

(ii) It can repair 600 sheets per day.

(iii) Life expectancy of machine is 20 years.

(iv) Rate of interest is 12% per annum.

To work out, the amount per sheet, the total amount in 20 years and total number of sheets Lamination in this period, must be known.

(a) Amount = Principal + Interest

= Rs. 10,00,000 + Rs. 24,00,000/-

= Rs. 34,00,000/-

(b) Number of Sheets Laminated in 20 years.

= 600 x 20 x 12 x 20 = 28,80,000

(No. of Sheets laminated in a day by a man x 20 days x 12 months x 20 years)

Amount per sheet = 34,00,000 ÷ 2,880,000 = Rs. 1.20

(5) Electricity Charges: If electric consumption to run lamination machine and for other purposes amounts to be Rs. 3000/- month, the electric expenditure per document can be worked out deviding the amount by the total number of sheets laminated during the month.

Expenditure per sheet = 3000 / 12000 = Rs. 0.25

[Number of sheets Laminated in a month = 600 x 20 = 12000 appr.]

Thus the total expenditure incurred for laminating A4 size document by machine can be found, summing expenditure towards items 1 + 2 + 3 + 4 +5

i.e. Rs. 7.66 + Rs. 3.40 + Rs. 0.95 + Rs. 1.20 + Rs. 0.25 = Rs. 13.46.

30% of 13.46 = Rs. 4.00.

Thus, the cost involved for laminating one A4 size document would be Rs. 17.50 - without any profit. If Re. 1.00 per sheet is clubbed as miscellaneous changes, this expenditure per sheet becomes Rs. 18.50 per sheet.

By making slight modification in the formula, the expenditure involved for preservation or for any reinforcement technique can be found out.

ANNEXURE-III

REQUIREMENT OF PAPER FOR DOCKET/FILE COVERS IN PRESERVATION OF RECORDS

Docket/file covers play an important role in Preservation of documents as they protect them against stress & strain caused during handling, storage & transportation. The paper to be used for the purpose should not only be free from acidity but also have permanent and durable character. According to Mr. Barrow, the good quality covering paper to be used for permanent records should have folding endurance not less than 1,000 double folds under one kg. tension and retention of this strength after accelerated ageing should not reduce below 45%. These are the basic requirement of good quality covering paper (docket/file cover) set 50 years back. Now the standard specifications for these materials according to category of records are available, depending upon retention period say (i) 100 years or more, (ii) 25-50 years, and (iii) 5 to 10 years.

Category I	
• Folding Endurance under one kg. tension	1000 d. folds
• Retention of folding endurance after ageing	70% or more
• PH value	Above 6.
• Copper number	Not more than one.
• Alpha cellulose content	More than 88%
Category II	
• Folding Endurance under 0.5 kg. tension	More than 300d. folds
• Retention of folding endurance after ageing	More than 50%

• PH value	5.5 or more
• Copper number	Not more than two
Category III	
• Folding Endurance at 0.5 kg tension	100 or more d. folds
• Retention in folding endurance after ageing	25% or above
• PH value	5.00 or more
• Copper number	Not more than two

Since the physical & chemical requirements of docket/file cover paper, supplied by stationery office Kolkata, to various creating agencies, custodial organisation and also to ministries was not up to the mark, the conservation Research Laboratory of National Archives of India, took up the study of materials in 1972. Under the scheme, cover papers received from manufacturing mills[+1] as well as from Hand Made Paper Manufacturing Units[+2] was tested and found that the PH range in case of mill paper was 5.5 while in case of Handmade paper it was above 6. Means the paper manufactured by Hand Made Paper Units was less acidic than the paper prepared by Mills. Depending on the life expectancy of records of different categories say, 100 years or above, 25-20 years and 5-10 years, it was concluded that the cover paper manufactured by handmade units is suitable for keeping records more than 100 years while for records of second & third category (to be retained 50 years & 10 years) the cover paper manufactured by mills can be used.

[+1] *Paper Manufacturing Mills*	[+2] *Hand made paper manufacturing units*
(i) M/s Bengal paper mills	(i) M/s Hand made paper Industries, Banglore.
(ii) M/s Orient paper mills	(ii) M/s Khadi Gram Udyog Bhavan, New Delhi & Kalpi.
(iii) M/s Rohtas Industries	(iii) M/s Hand made paper centre, Shri Aurobindo Ashram, Pondicherry.
(iv) M/s Sirpur paper mills	(iv) Hand made paper Institute, Pune.
(v) M/s Titagarh paper mills.	

ANNEXURE-IV

STANDARD SPECIFICATIONS OF MATERIALS

The standard specifications of various stationery materials for ensuring quality of materials to be used in archival conservation are given below.

(i) Handmade Paper

(i) Thickness- .0033" to .004" (making a subtance 9-10 kg. per ream).

(ii) Subtance G.S.M.-59 gm/M^2.

(iii) Folding Endurance under one kg tension—

(a) Unaged not less than -500 double folds.

(b) Aged not less than 400 double folds.

(iv) Retention in folding-endurance 80% or above.

(v) PH—(a) Unaged—not less than 6.5

(b) Aged—not less than 6.00

Note: After ageing, the paper should not show any adverse change incolour.

PH of paper above 7 & below 9 will be more regarded.

(ii) Paper for docket/file covers

Specifications for covering material for A class records to be retained permanently.

(i)	Thickness	.0304 to .0405 cm.
(ii)	Substance-G.S.M.	180-250 G.S.M.
(iii)	Folding Endurance under-one kg tension	Not less than 1000 double folds.

(iv) Retention of folding Endurance after ageing	70% or above.
(v) PH	Not less 6.5 (Better would be if PH is more than 7)
(vi) Alphacellulose content	More than 90%.
(vii) Copper number	Not more than 0.25.
(viii) Colour	Fast to ageing, water & alcohol.

(iii) Bond Paper

No standard specification are available. However, tentatively, a standard bond paper for archival use should conform as under:

(i) Substance (G.S.M.)	Not less than 70 GSM.
(ii) Folding endurance- Under one kg tension	(a) Not less than 100 double folds unaged. (b) Not less than 70.
(iii) Retention in folding- Endurance	70% or more. (above 60% may be compromised).
(iv) PH-	(a) Unaged more than 6.5 (b) Aged not less than 6.00 (Better would be if PH is more than 7)
(vi) Colour	Fast to ageing, water and alcohol.

(iv) Ledger Paper

(i) Folding Endurance under 0.5 kg tension.	Not less than 200 double folds.
(ii) Percentage retention in Folding endurance	Not less than 60%.
(iii) PH	(a) Not less than 6.5 unaged (b) Not less than 6.20 aged. (Better would be if PH is more than 7)

(v) Tissue Paper

(i) Thickness-	0.2 mm.
(ii) Substance G.S.M.	10-14 G.S.M.
(iii) PH-	Above 7 (unaged) & not less than 7 (aged).
(iv) Alphacellulose content	More than 88%.
(v) Ash content	Not more than 0.5%
(vi) Colour	White fast against ageing, water and alcohol.

(vi) Blotting Paper

Colourless Paper, with PH value not below 5 having good water absorbancy is considered a good quality blotting paper for archival use.

(vii) Boards

The binding boards, made of single ply, uniform thickness, hard possessed, flat with even surface, clean & sides free from loose fibres are considered quality boards for binding. However, characteristic requirements of straw board & Mill board are as under.

(i) *Straw Board*

Thickness	*PH.*	*Substance G.S.M.*	*Ash content*
Less than 1.5 mm	Not less than 6.5	1000 G.S.M.	Not more than 15%
Less than 3.5	-do-	2000 G.S.M.	-do-
Less than 4.5	-do-	3000 G.S.M.	-do-
(ii) *Mill Board*			
Less than 1.5	Not less than 6.5	1500 G.S.M.	Not more than 10%
Less than 3.5	-do-	3500 G.S.M.	-do-
Less than 4.5	-do-	4500 G.S.M.	-do-

Note: Better would be if PH of any kind of board is above 7.

(viii) Covering Materidh

Art canvas, Rexine & binding cloth (calico) are generally used for covering of binding boards.

Characteristic Requirements

Type of cloth	*Weight of the finished Fabric gms/ Sq.m.*	*Percentage of coating & filling over basic weigh*	*Thread count average warp and filling (2.5 × 2.5 cm)*	*Ash content (% of finished cloth)*
Art canvas-	Between 400-800	less than 35	70 double threads	less than 5
Calico-	Between 150-250	less than 70	110	less than 15
Rexene	Between 250-450	less than 80	120	less than 15

(ix) Leather

Goat or Sheep

Leather, vegetable tanned or chrome tanned or combination of the two, free from grain defect or insect damage, conforming to IS: 2960-1964, [Specifications for Book binding-Leather] is suitable leather for quality binding.

(x) Reinforcement Fabrics [Long Cloth]

The long cloth is used on the spine, folds of end paper and for map mounting etc. The unsized cloth, free from weaving defects & loose threads is considered good fabric for quality reinforcement.

Characteristic Requirements

Type of Cloth	*Thread count warp & filling average (2.5 × 2.5 cm).*	*Thickness average*	*Weight gm/sq.M.*
Long cloth	160	0.15 mm	120

(xi) Sewing Thread

Thread made of unbleached linen or cotton, not less than 3 cords, unsized, capable to withstand breaking load of not less than 2.5 kg. is good material for sewing of Book while for reference volumes, it should be capable to withstand breaking load of 5.00 kg.

ANNEXURE-V

DISASTER PLANNING IN ARCHIVES & LIBRARIES

Disaster planning is an accepted part of Archives & Library preservation in most of the developed countries while in developing and under developed countries it is still to take a ride. No account of loss brought out by disaster elements in India is available but quantum of loss occurred due to these in various countries of world is an eye opener. For instance, 400,000 books in Los Angles Public Library in 1986, 400,000 vols and news paper collections at library of the USSR academy of sciences were destroyed by the arson.

Among the floods and Hurricanes, few are the incidents occurred in Biloxi; Mississippi (in 1968), and the University of Corpus Christe Library (in 1970). The Cunnecticut River overflowed into the cunnecticut historical society library in 1955 and destroyed unaccountable number of library collection beyond recovery. Similarly, flood occurrences of Gulbenkian Museum near Lisbon (1967), Fondren Library at Rice University (1966) and Rockport Historical Society destroyed innumerable books, vols and other materials.

So far elements of disaster, they are not only the flood and fire but earthquake, land sliding, war in general and atomic war in particular, terrorist activities and cyclones (tornados and hurricanes) also make part of.

Once fire out bursts and goes ghostic, leaves nothing but half burnt mass, charcoal and ash. Floods occurred because of any reason, washout information recorded on paper with water soluble inks and convert paper into pulp if not taken care of properly.

When hurricanes sweep in from the sea or rivers flood their banks, building on low-lying land are inundated and suffer broken windows and collapsed walls and roofs if not storm proof. Books, volumes and files under tons of mud and building debris could start to mold as soon as the water and wind subsides.

Land sliding smashes building structure, carries the housed wealth down the steeps buried with soil and water currents take them away untraceable. Earthquakes on the other, burry the wealth under material used for building construction. Inexperienced recovery team, without sense of archives, put them into more losses than brought out by the incident.

Wartime happenings bring about losses to the wealth by one way or another. Atomic wars, however, pose more serious problem to the heritage including building structure. Experiences of Ist and IInd World Wars say that this wealth was safer in Archives and Libraries located in small townships than metropolitan cities, near, mines and military cantonments. Obviously, the damages caused by any of the disasters are serious and need disaster planning in Archive and libraries if documentary patrimony is to be preserved for posterity.

Disaster Planning

Whatever, infrastructure about disaster planning is available today world over is not the outcome of a research and development. But it has been achieved in order to counteract the problems faced by the people of Archive and libraries because of disaster occurrences. What I understand is that disaster management needs a meticulous planning including the basic question of suitable archival and library building and its location. In this respect following points in general should be well considered by those who are serious about the preservation of wealth against disasters.

(i) *The Cite*: near the bank of a river and nalah, the area prone to frequent earthquake and land sliding, is not considered a good location to have a building for archives and libraries. If any exigencies to have such building arise, the suitable changes in the building structure plan and materials to be deployed for construction should be considered carefully in the light of disaster occurrences.

(ii) The modern technology to have an earthquake proof building should be employed to the maximum possible extent in true senses. The recent earthquake in Gujarat (2000) and other parts of the country and the losses to the building structures are more than sufficient to convince the people about importance of earthquake proof buildings today.

(iii) The modern building structure and material (steel and concrete) have been found unsuitable to protect the housed wealth against atomic wars. The reason behind is that heat generated by an atomic bomb, softens the steel, causes girders to buckle following fall of one floor successively on the other. The old stone arch buildings, however, does not collapse easily.

(iv) Buildings making use of glasses in huge number or glass palaces are not the suitable buildings for housing documents and books, in tornado and hurricane prone areas. Installation of shutters at the first alert may considerably prevent major losses by wind and water. Texas library has reaped a rich reward of this planning.

(v) A building having fire proof character and with facilities of fire detection and combating, is considered suitable for preserving documentary heritage against arson. Besides, information on the following by every Archives and library should make ready to deal with disaster occurrences.

(a) A list of arrangement for regular inspection of security equipments such as fire detection alarm system and fire extinguishing equipments.

(b) List of arrangements for building inspection covering building maintenance, roofs, drainage water pipes, plumbing and electrical facilities.

(c) List of local firms with their telephone numbers, dealing with fire fighting equipments.

(d) Floor plans of the building with location of priority materials and salvage equipments.

To meet out the challenges posed by any disaster, a four fold planning with elements of *prevention, response, reaction* and *recovery* have got favour of various organizations of world today.

Prevention

A famous proverb, "Prevention is better than cure", itself advocates the importance of prevention than other activities of post occurrences. A meticulously planned, preventive programme, minimizes the damage which a disaster will cause. The features of such planning include the making of security copies of important materials, regular inspection of materials housed and building, maintenance and house keeping of wealth. About the building, however, information on the following should be gathered and action should be initiated well in time.

(i) Age of building and its life expectancy.

(ii) Whether or not any fire or water incident occurred? If so, whether the causes for such occurrences eliminated or temporarily patched up.

(iii) Are the windows tight or they left open for ventilation?

(iv) Are all the heating plants protected with fire doors?

(v) Is the fuel storage safe?

(vi) Is the insulation material of electrical wiring safe or become worn or deteriorated?

(vii) Are the roofs of the building leak proof?

More to be added is that regular instructions to staff in fire-prevention process, maintenance of fire detecting and fire fighting equipments and preparedness of staff to attend to such activities, minimizes the possibility of out break of fire and spread.

Response

The second phase of disaster planning is response. The major steps involved towards this element are the constitution of a response team and training. In the hours of peace, the team taking initiative, should effort to get the needful done on every point mentioned in prevention phase. In other words, the team should ensure that the required information has been collected, security copies are made and pending work on other items has been completed well in time. Besides, the response team should also collect the information about the availability of conservation personnel and the facilities in the region. If need be, a contract may also be reached to make the personnel available and facilities in the post disaster happening. Arrangement for access to freezer capacities and mechanism to pay for emergency needs is not less important and the response team should complete it. Any other thing, thought to be useful for preventing disaster and to meet the challenges of post occurrence should also be incorporated in the list of activities to be performed by the response team. The disaster plan, regularly updated, is to be more effective in disaster planning.

Reaction

This phase of disaster planning is concerned with the procedures to be followed in the post occurrence of disaster. The first step in this respect is of raising alarm to assemble the personnel. The second step is the safer escape of staff and the third is to make assessment of disaster site whether or not it is safer to enter. The activity is followed by the estimation, preliminary, about the damage. It is the joint activity of disaster team but the leader of the team is supposed to take initiative without loss of time. Once the extent of damage has been estimated, the suitable steps are taken for their quick and safe removal, packing, transporting and storage in a safer area. Sometimes tents are made for storage of removed material. In the meanwhile, a treatment centre is set up close to the storage area and equipments and materials required for treatments are collected to carry out minor treatment of material, in consultation with expert conservationists.

Salvage

All the efforts of disaster recovery team to remove the materials from disaster cite will go invain if proper salvaging facility is not provided to disaster hit records specially with water. Air-drying treatment to slightly wet records is the best way of salvaging. The principle behind the technique is to expose the surfaces of documents to the open environment because of which quick vapourization of water takes place lowering the water contents and drying the material. Blowing of air by air circulators or by other artificial means, hastens the drying process. Alternative is to interleave the pages with blotting or toweling papers and replace them frequently as often as necessary until dry. The drying technique, theoretically, looks to be very easy but practically very difficult. Few questions to carry out the treatment safely divert the attention to be considered by the people and they are related to covered space for treatment, equipments for treatment, manpower to handle the work, the weather and atmospheric conditions of the treatment site and the supply position of electrical power energy. These are the salvaging aids without which the pace of treatment would be very slow and sometimes ineffective.

Completely soaked records in masses, however, cannot be saved by the air-dry technique. For, their treatment a newly developed system called Freeze-drying technique is found very suitable. The technique was invented in 1972. The completely, soaked records or organic carriers are frozen usually in a commercially blast freezer at temperature of -21°C or below. The small lot of frozen records is subsequently subjected in a vacuum chamber. The reduced pressure of the chamber brings down the boiling point of water causing it to evaporate without going through the liquid state. The vapours are drawn off systematically. The remaining wet frozen lot is thus treated in piecemeal and dried successively.

Air purging is the another device for treating completely wet records in mass scale. In this technique whole of the building or site housing wet material is sealed by wrapping plastic sheets and air loaded with temperature at 26°C and 15% relative humidity is pumped through.

Disinfection

Wet record materials partially or fully, in general invite fungus or mould infestation. As such, mould control techniques are often required in after math of disaster. All the records affected with water, partially or fully, for the purpose should be given disinfestation treatment by 10% solution of thymol in methylated spirit as a preventive step against botanical infestation. This treatment is repeated after every fortnight especially in humid climatic conditions. Simultaneously, the examination of records should be carried out and the infested records and the records doubted to be attacked by the agency are segregated and treated as under.

(i) Any mould present is cleaned off mechanically and the area is treated with 5-10% solution of thymol solution in methylated spirit. Subsequently, pages are interleaved with thymolised paper.

The books on Art paper i.e. the glossy paper are ruined by immersion in water and therefore it would be wiser step to give priority treatment to such materials. A bound book or volume under treatment should never be opened more than a finger's breath as wide opening at early stage may damage its opine.

The treatment of wet documentary material with CO_2 (carbon dioxide) gas in vacuum fumigation chamber minimizes the chances of mould growth. Gamma irradiation technique is also suitable for killing mould cells. It is an effective non-chemical technique to destroy moulds but it has adverse effect on cellulose and proteins.

Last but not the least, Archives and Libraries without trained staff, to deal with the material in the odd hours, whatever facilities they have, can not do justice with the wealth and hardly can reap a rich reward. As such trained staff is necessity of every custodial department for disaster management.

BIBLIOGRAPHY

Agarwal, O.P., "Environment & Preservation of Paper", *Conservation of Cultural Property in India, Vol. X, 1977*, p. 15.

---------, "Problems of preservation of Indian Murals," *Conservation in the Tropics, 1972*, pp. 87-90.

---------, Suryavanshi, D.G. and Gupta C.B., "Effect of Movilith on Fibrous Materials", *Conservation of Cultural Property in India, Vol. XVI & XVII 1983 & 1984.*

---------, & Bisht, A.S., "Non-Aqueous Deacidification and Conservation of an Indian Illustrated Manuscript", *Studies in Muscology, Vol. V-1969* pp. 34-37.

Alexandra, Soteriou, *Gift of Conquerors, Hand Paper Making in India.*

Baji, A.N., "Some Problems Relating to the Preservation of Marathi Documents", *The Indian Archives, Vol. XII, No. 1-2, 1958*, p. 76.

Basu, Purendu, "Why Preservation", *The Indian Archives, Vol. No. 1- 1947*, p. 20.

---------, "Enemies of Records", *The Indian Archives, Vol. IV, No. 1, 1950*, pp. 7-13.

Bell, H.E., "Protection of Archives, Some Lesson from the War", *The Indian Archives, Vol. IV, No. 2, 1950*, pp. 164-66.

Bhandari, K.S. and Srivastava, Anita, "Wet Cleaning of Straws for Improving Pulp Quality", *IPPTA, Vol. 10, No. 3, July-Sept. 1998*, p. 159.

Bist, A.S., "Preservation of An Illustrated Manuscript on Paper", *Conservation in the Tropic, edited by O.P. Aggarwal, 1972*, pp. 98-102.

Borne, Esterk, "Archives and War", *The Indian Archives, Vol. IV, No. 2, 1950*, pp. 150-59.

Buck, E.A., "Bookworms", *The Indian Archives, Vol. 1, No. 2, 1947*, pp. 126-34.

Chakraborty, S. and Manjumdar, P.C., "A note on Sulph-Arsenic", *The Indian Archives, Vol. No. 2, 1947*, pp. 114-17.

---------, "The Effect of Gamma XENIX on Paper", *The Indian Archives, Vol. II, 1948*, p. 55.

---------, "A Review of the Lamination Process", *The Indian Archives, Vol. I,*

No. 4, 1954, p. 38.

Chand, Punam, "Control of Termites" *Conservation of Cultural Property in India, Vol. IX, 1976*, p. 12.

Conservation of Traditional Records, "Paper and Allied Materials", *National Archives of India*, p. 40.

Cultural Heritage of India, Vol. III, Ramkrishna Mission, 1953.

Das, Amal Chandra, "Adhesives", *Conservation of Cultural Property in India, Vol. XVII, 1983-1984*, p. 90.

Das, Sudhir Kumar, "Repair & Preservation of Water Damaged Documents", *Conservation of Cultural Property in India, Vol. VII, 1979*, p. 43.

Davis, John, "Preventive Conservation of Archives", *Conservation of Traditional Records Paper and Allied Mateirals*, pp. 71-77.

---------, "Polyester film Encapsulation, A Restoration Technique", *Conservation of Traditional Records, Paper and Allied Material*, p. 166.

Dutta, P.K., "Conservation-The Potentialities of Neem Oil for Controlling Insect Attack", *Conservation of Cultural Property News Letter, No. 14, 1986* p. 7.

Dutt, Hirendra Nath, *The Twentieth Century Cyclopedia of Ink and Allied Proct, Vol. 1*, p. 4.

Ghose, Arun, "Note on the Classification of Choice of Adhesives" *Conservation of Cultural Property in India, Vol. XV, 1993*, p. 18.

Glossary of *Keeping Archives*, Published by Australian Society of Archives Sydney, 1987.

Goel, O.P., "Repair of Documents with Cellulose Acetate Foils on Small Scale", *The Indian Archives, Vol. No. 2, 1953*, p. 162.

Grant, Julius, "Analytical Methods in the Dating of Books and Documents", *The Indian Archives, Vol. I, 1947*, p. 36.

Gupta, C.B., "General Conservation Problems, in Tagores Art on Pager", *Conservation of Cultural Property in India, Vol. X. 1977*, p. 54.

Gupta, Motilal, "Difficulties found in the Collecting and Conserving Cultural Materials, Manuscripts," *Conservation of Cultural Property in India, Vol. XII, 1979*, p. 54.

Gupta, R.C., "Building and Storage Equipment Congenial to Long Term Preservation of Paper", *Conservation of Cultural Property in India, Vol. X, 1977*, pp. 29-32.

--------, "How to Fight White Ants", *The Indian Archives, Vol. III, No. 2, July-Sept. 1954*, pp. 120-22.

--------, "Effect of Clenofix and Other Common Insecticides on the Durability of Paper", *The Indian Archives, Vol. XVI, December 1966* pp. 29-34.

--------, "A Review of Work in Conservation in the National Archives of India", *Conservation of Cultural Property in India, No. 1, 1966* pp. 54-61.

---------, Effects of Alternative Heating and Colling on Physical Properties of Paper," *The Indian Archives, Vol. XIX, No. 2, July-Dec. 1970,* pp. 25-27.

Gunther, Richard, "The Durability of Paper", *The Indian Archives, Vol. 1, No. 3, 1947,* p. 231.

Harold, S. Burt, "Inks and Papers for use in Govt. Officers in U.S.A.", *the Indian Archives, Jan-Dec. Vol. X, Nos. 1-2, 1956,* p. 13.

Hotchand, "Ecosystem & Fungal Growth" *Conservation of Cultural Property News Letter, No. 15, January 1988* p. 9.

Hughes, Mckenny A.W., "Protecting of Books and Records for Insects", *The Indian Archives, Vol. VI, No. 1, 1953,* pp. 41-43.

Indian Forest Bulletin No. 127, Cellulose & Paper, 1957, pp. 1-3.

IS: 1774-1986 (Reaffirmed 1990), Standard Paper for Permanent Records.

IS: 220-1988, Ink for Permanent Records.

IS: 1581-1975, Ink for Permanent Records.

IS : 5805-1993, Specification for Ball Point Pen Inks.

Indian Historical Records Commission, A Retrospect, Vol. II, 1948-1994, p. 10.

IS: 1551-1959, Carbon Paper for Typewriters.

IS: 41-74-1967, Typewriter Ribbon Ink.

Indian Historical Records Commission, A Retrospect, (1919-1948), Vol. 1, p. 3.

Indian Historical Records Commission Proceedings, Vol. 28, 1951.

Ibid., Vol. 36, 1961.

Ibid., Vol. 40, 1970.

Ibid., Vol. 41, 1971.

Ibid., Vol. 42, 1973.

Ibid., Vol. 43, 1975.

Indian Historical Records, Vol. 46, 1979.

---------, *Vol. 49, 1985.*

---------, *Vol. 50, 1987.*

---------, *Vol. 52, 1988.*

---------, *Vol. 53, 1990.*

IS: 2663-1964, Indian Standards Code of Practice Relating to Primary Elements in Design of Building of Archives.

Indian Historical Records Commission, A Retrospect, 1948 to 1994, Vol. II.

Joshi, M.C. and Dey, *Ink, Paper and Art of Writing in Perspective of Indian Antiquity,* International Seminar on Examination of Documents held in Madras, 1997.

Kesavan, B.S., *India's National Library,* p. 49.

Kirk RTF, *Paper Making in the Bombay Presidency,* A Monography.

K.K. Guha Roy, "Library of the Indian Agricultural Research Institute", *The Indian Archives, Vol. 1, No. 2, 1947,* p. 122.

Kathpalia, Y.P., "*Restoration of Document*".

--------, Synthetic Fibres for Preservation of Paper, A Comparative Study", *Conservation of Cultural Property in India, Vol. 1, 1966*, pp. 62-73.

--------, "Sterilization of Records Against Fungus Insects & Rodents", *Conservation of Traditional Records Paper and Allied Materials*, pp. 94-97.

Kauta, B.B., and Bhagwat, V.W., "Effect of Some Metalions on Degradation of Cellulose in Alkaline media", *IPPTA Vol. 10, No. 4, Oct. Dec. 1998*, p. 77.

Kimberly, Arthur E., "Recent Development in Records Preservation". *The Indian Archives, Vol. III, No. 1-4, 1949*, p. 70.

Kishore, Ranbir, "Housing Requirements of Conservation of Archives", *Conservation of Traditions Records Paper and Allied Materials* pp. 78-80.

Lal, Avinashi, "Destructive & Non Destructive Methods of Analysis", *Conservation of Cultural Property in India, Vol. VI, 1978*, p. 54.

--------, "Restoration of Fragile Documents", *Conservation of Traditional Records Paper and Allied Materials*, p. 111.

Lal, B.B., "Scientific Examination of Documents, Works of Arts, Monuments and other Cultural Property in India", *Conservation of Cultural Property in India, Vol. VI, 1978*, p. 21.

________"Some Conservation Problem in Kathmandu, Nepal, with Special Reference of Preservation of Bhaktpur Murals", *Conservation of Cultural Property in India, Vol. IV, 1967*, p. 69.

Mahajan, Shusma, "Upgradation of Wheat Straw and Rice Straw for Pulp & Paper Making", *Indian Pulp & Paper Technical Association, Vol. 10, No. 3, July-Sept. 1998*, p. 93.

Mehra, C.P., "Use of Naphthalene as Fumigant for Books and Manuscripts in Libraries and Record Repositories", *The Indian Archives, Vol. VIII, No. 2, July-Dec. 1954*, pp. 164-65.

Mitra, R. Lala, "The Report on the Manuscripts in India Repositories", *Indian Archives, Vol. II, 1948*, pp. 21-23.

Nair, S.M., "Biodeterioration of Museum Materials in Tropical Countries", *Conservation in the Topics, 1972*, pp. 150-58.

______"Biodegradion of Paper", *Conservation of Cultural Property in India*, Vol. X, *1977*, p. 13.

--------, "Some Simple Methods for Examination of Organic Museum Objects for Insect Attack", *Conservation of Cultural Property in India*, Vol. XI, *1978*, pp. 2-29.

"News Notes on Preservation of Archives", *The Indian Archives, Vol. IV, No. 2, 1950*, pp. 242-43.

Panigrahi, J.C. and Patel, M., "Stock Preparation Strategies in Quality Paper Manufacturing-Part, Normal & Reverse Sizing", *Indian Pulp & Paper Technical Association, Vol. No. 1, Jan.-March 1998*, p. 35.

Paul, S.K. & Colleagues, "Neutralsizing of Bagasse Furnish-A laboratory Study", *Indian Pulp & Paper Technical Association, Vol. No. 3, Sept. 1997,* pp. 61-67.

"Postwar, Reorganization of Archives Office in India", *Indian Historical Records Commission, Vol. 21,* Dec. 1944.

Prajapati, C.L., *Archivo Library Material, Their Enemies and the Need of First Phase Conservation,* 1997.

---------, "Conservation of MSS. In Tawny Monastery", *The Indian Archives Vol. XXXIX, No. 1, Jan.-June 1990,* pp. 23-28.

---------, "Tracing History of Growth of Conservation and its Development in National Archives of India", *The Indian Archives,* Vol. XLVI, Nos. 1-2, Jan.-Dec. 1997, p. 54.

---------, "New Trends in Conservation", *Proceeding of the Meeting National Committee on Archives held at Shimla, 1993.*

---------, "Conservation of Endangered Documents", *Proceedings of the Meeting of NCA, held at Jaipur, 1998.*

---------, "Problems relating to Establishment of Effective Preservation & Restoration in Archives and Libraries", *Lucknow Librarian, Vol. 27, Nos. 1-4, 1995,* p. 37.

---------, *Modern Techniques on Conservation of Palm Leaf Manuscripts, and Other Manuscripts in Indian Languages,* p. 201.

---------, "Elements of Records Management & Conservation", *National Archives of India,* New Delhi, 1993.

---------, *Pralekhon. Pandulipion Tatha Dusprapya Pustkon Ka Parirakshan & Prati Sanskar (In Hindi).*

Prasad, Rajindera, "High Grade Paper for Record and other Printed Materials of Archival Interest", *Conservation of Cultural Property in India, Vol. IX, 1976,* p. 34.

Prasad, S.N., "Some Observations on Conservation of Archival Material", *Conservation of Cultural Property in India, Vol. VI, 1973,* p. 11.

Proceeding of the 10th Meeting of National Committee on Archives, Simla, 6-7, July 1956, p. 3.

Rajesh, K.S. and Rao, Mohan N.R., "Bagasse the Promising Alternative for the Future", *Indian Pulp and Paper Technical Association, Vol. 10, No. 3, July-Sept. 1998,* pp. 151-57.

Rao, A.R.K., Bhattacharya, J.P. & Benerjee, Saugata, "Current Status of Indian Paper Industry and Outlook for Investment in Near Future", *Indian Pulp & Paper Technical Association, Vol. 10, No. 4, Oct.-Dec. 1998.*

Rao, Y.K., *The Paper Industry in India, Status and Prospects.*

Rawling, G.I.F., "The National Gallery in War Time", *The Indian Archives, Vol. IV, No. 1, 1950,* pp. 24-34.

Roonwal, M.L. and Chatterjee, P.N., "The Control of the Indian Bookworm Beetles. Gastrallus Indicus Reitter", *Indian Forest Records, Entomology Vol. 8, No. 6, 1952,* pp. 53-54.

Roy, Guha, K.K., "*The Indian Archives*", *Vol. 1, No. 2, 1947*, p. 124.

Sarkar, N.N., "A New Sterilisation Process and Standardisation Process and Standardisation of Lethal Dose for Insects Eggs etc. by Vacuum Fumigation Chamber", *Conservation of Cultural Property News Letter, No. 14, November 1986*, p. 15.

Satyendra, *Pandulipi Mala.*

Schol, Tamas, "New Method of Determining the Surface pH of Paper by Plane Class Electrode", *The Indian Archives, Vol. XXI, No. 1972*, p. 51.

Sen, S.N., "Common Errors", *The Indian Archives, Vol. 1, No. 4, 1954*, p. 304.

Shaheen, Fauzia, "The Insect Factor in Deterioration of Wooden Cultural Property", *Conservation of Cultural Property in India, News Letter No. 15, January 1980*, p. 12.

Sharma, Chhaya, Rao, M.J. and Singh, S.P., "Effect of Recycling of Fibre *Characteristics" Indian Pulp & Paper Technical Associated (IPPTA), Conservation Issue, December 1998*, pp. 69-73.

Sharma, S.K., "Scientific Examination of Documents", *Conservation of Cultural Property in India, Vol. VI, 1978*, p. 31.

Sinha, S.N., "Relevance of Conservation of Archives", *Conservation of Cultural Property News Letter, No. 15, January 1988*, p. II.

Status Report on Resolution Passed by the Indian Historical Records Commission (40th-52nd Session), Rohtak, 20-21 Apr., 92.

Steckzen, Birger, "Storage & Preservation of Maps in Swedish Military Archives", *the Indian Archives, Vol. IV, No. 1, 1950*, pp. 14-19.

Talwar, V.V., "Fumigation Techniques for Sterilisation of Paper", *Conservation of Cultural Property in India, Vol. VIII, 1975*, pp. 46-48.

---------, "Loading & Filling", *Conservation of Traditional Records, Materials*, pp. 68.

---------, "Trends in Pest Control", *Indian Archives, Vol. XLII, No. 1-2, 1993*, pp. 129-131.

---------, "Care of Water & Fire Damaged Records", *Conservation of Cultural Property, in India, Vol. X, 1997*, p. 40.

Tandon B.N., "Problems in Conservation of Miniature Paper Paintings and Illustrated Documents", *Conservation of Cultural Property in India, Vol. X, 1977*, p. 38.

Thomas, M. Iiams, "Preservation of Books & Manuscripts", *The Indian Archives, Vol. 1, No. 1, 1947*, p. 39.

Utillisation (News Service) Indian Forest Bulletin, No. 108, 1941, pp. 1-10.

INDEX

PLATES

1.Beating of Pulp with dhegis (Ahmedabad)

2. Taking out wet films on a screen (Ahmedabad)

3. Ruined Chabutras used at Erandol-till 1908 AD

4. Washing Paper Pulp in a Dhoti, at the river side in Erandol

5. Drying Paper sheets on the wall (Ahmedabad)

6. Experimental mechanical pulp grinder, installed in Forest Research Institute, Dehradun 1937

7 (a). Indigenous machine for producing strawboard

7 (b). Indigenous machine for producing strawboard

8 (a). Indigenous machines (Digester and Beater)

8 (b). Indigenous machines (Digester and Beater)

9 (a). Indigenous machines (steam heated drier and cross cutter) for producing Straw Board

9 (b). Indigenous machines (steam heated drier and cross cutter) for producing Straw Board

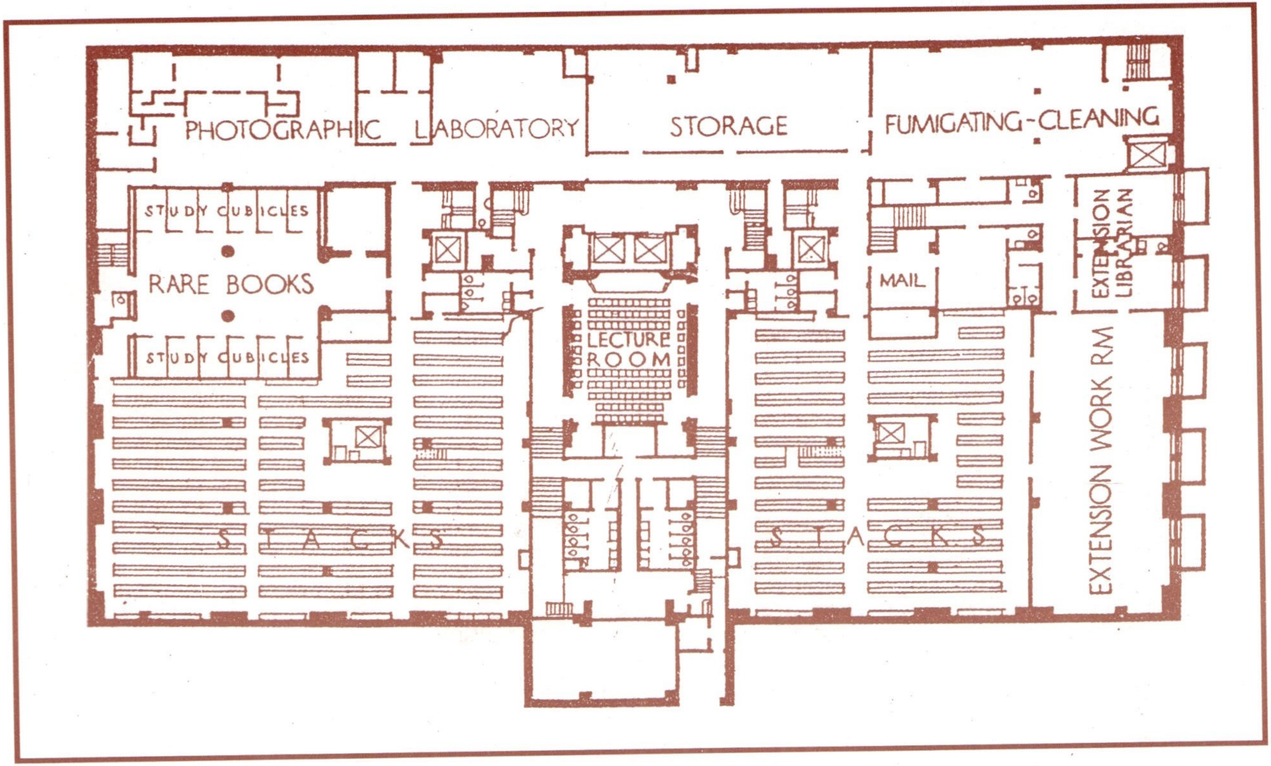

10. Basement plan Adopted in Virginia State Library

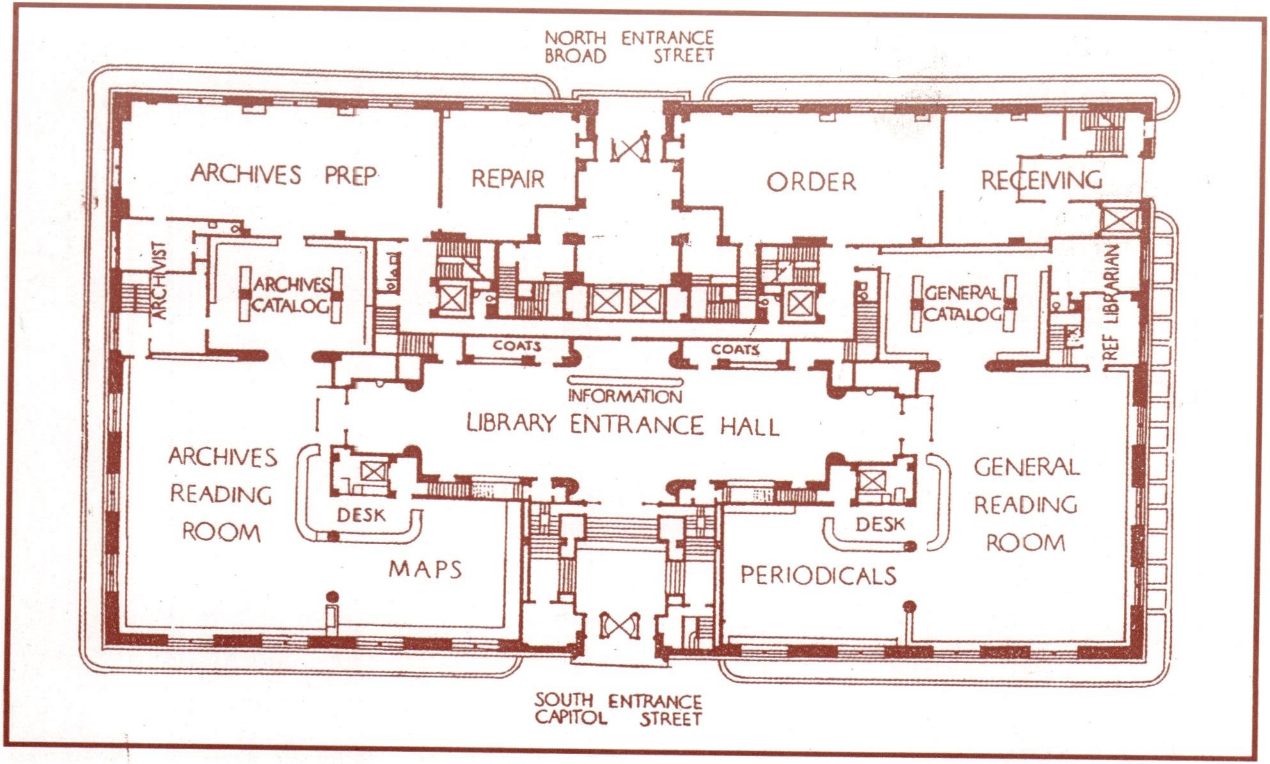

11. First Floor Plan 128' x 223' adopted in Virginia State Library

12. General view of the Stack Area, first floor, NAI

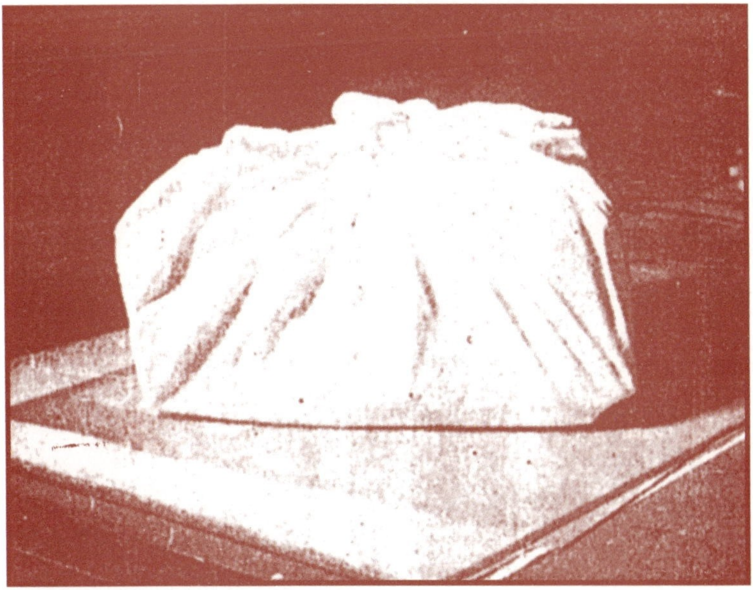

13. Storage in Bastas

18. The main underground approach level

19. A typical plant room and part of storage building

20. Part of storage chamber before occupation showing stack of screens

21. Inside one of the storage buildings

22. Closed Trolley (one of the four side-doors removed)

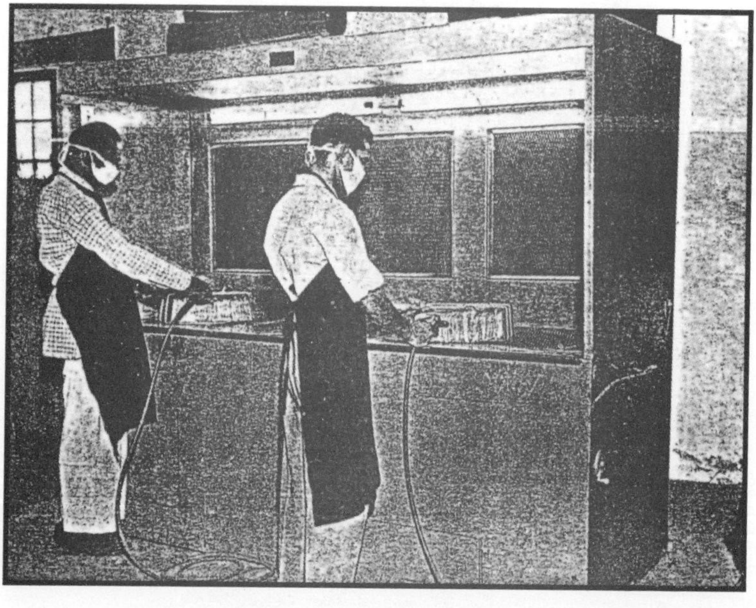

23. Air Cleaning equipment being used in NAI

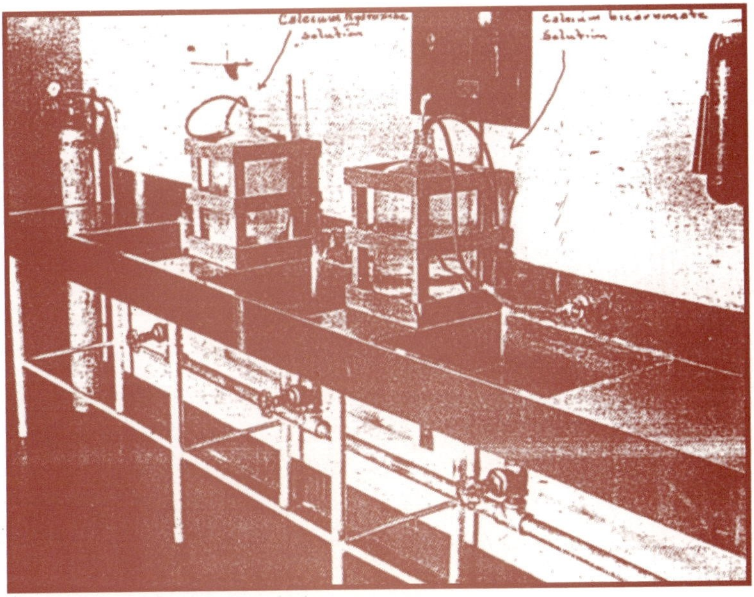

24. Arrangement for deacidification of documents using Barrow's technique

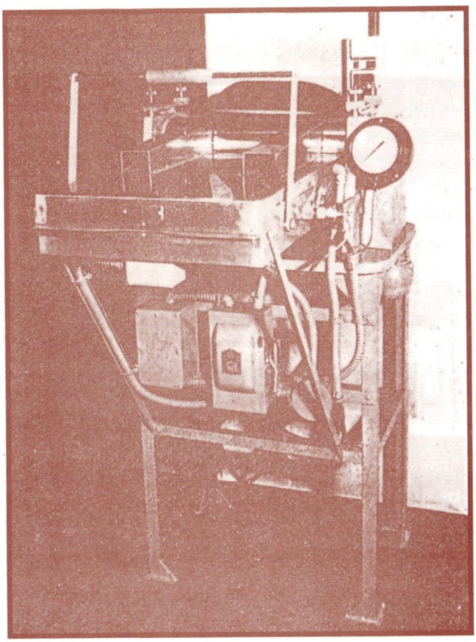

25. Barrow's Lamination machine

26. Flat-bed hydraulic laminator being used in NAI

27. Impregnator Laminator

28. Showing unsystematic storage of MSS in one of the Satras of Majuli

29. Showing Systematic Storage of MSS in Almirahs
in one of the Satras of Majuli

An illustrated leaf showing physical loss at the edges

The reverse of the same leaf showing charring and perforations exactly at the back of the particular green pigment.

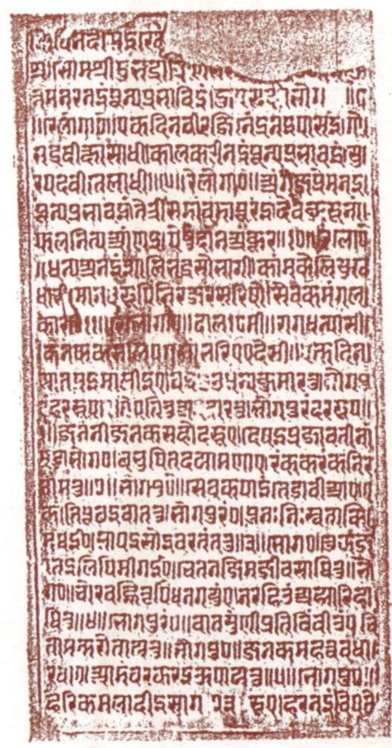

30. Showing Illustrared manuscripts